The Number Systems and Operations of Arithmetic

The Number Systems and Operations of Arithmetic

An explanation of the fundamental principles of mathematics which underlie the understanding and use of arithmetic, designed for in-service training of elementary school teachers and pre-service training of elementary school teacher candidates

ORVAL M. KLOSE

Professor of Mathematics, Humboldt State College,
Arcata, California

PERGAMON PRESS

OXFORD · LONDON · EDINBURGH · NEW YORK
TORONTO · PARIS · BRAUNSCHWEIG

Pergamon Press Ltd., Headington Hill Hall, Oxford
4 & 5 Fitzroy Square, London W.1.
Pergamon Press (Scotland) Ltd., 2 & 3 Teviot Place, Edinburgh 1
Pergamon of Canada, Ltd., 6 Adelaide Street East, Toronto, Ontario
Pergamon Press Inc., 44-01 21st Street, Long Island City, New York 11101
Pergamon Press S.A.R.L., 24 rue des Écoles, Paris 5e
Friedr. Vieweg & Sohn, Verlag, Postfach 185, 33 Braunschweig, West Germany

---

First edition 1966
Library of Congress Catalog Card No. 65-26342

Printed in Great Britain by Billing & Sons Limited, Guildford and London

(2422/66)

CONTENTS

PREFACE

THIS book was written for the single purpose of explaining to elementary school teachers (both in-service and in-training) the nature of those basic principles of mathematics which form the foundations and structural framework of arithmetic, and how the familiar formal algorithms of arithmetic stem from these structural principles.

The author has tried to present these ideas thru a frankly intuitive approach because on the basis of his experience he believes this to be the natural and best approach for those who teach elementary school children. It has been observed that these teachers are usually not impressed by rigorous axiomatic developments of theory and frequently do not share the enthusiasm for mathematical rigor in general which one might find in an advanced student specializing in mathematics. It is hoped and believed that one who follows the intuitive development presented here will culminate his experience in the formation of a sound understanding and user's appreciation of the basic concepts and principles underlying arithmetic and its teaching. If so, it is believed that such a common-sense understanding will make a more effective contribution to the teaching of arithmetic than would result from a premature emphasis on axiomatics.

Altho some of the material has been well received by secondary teachers of the author's acquaintance—and all those who find anything whatever in the book that is of value or interest to them are certainly welcome to make use of it—no claim is made for the book as a cure-all, with or without starred sections and exercises. As a matter of fact, on the basis of a belief that each competent instructor who uses the book will have his own good ideas on which topics should be included to provide challenge for his better students and which should be excluded to temper the wind to his less hardy lambs, no sections and no exercises have been starred. It will probably be generally agreed that most of the material in the first four chapters of Part 1 and most of the algorithmic material in Part 2 should be covered (say in a two semester sequence of courses meeting three hours per week); it will likewise probably be obvious to all that the material in Chapter 1.5 on abstract systems, and possibly also that on the complex number system at the end of Chapter 1.4, is in the challenge-and-enrichment department.

The author strongly believes that the most effective learning experience that can be stimulated thru the use of exercises is that in which the student first asks himself whether or not the results he has obtained are correct, pertinent and productive; also that the next-best situation is the one in which the student discusses these matters with his instructor. As a special favor to the student, and particularly that type of student (which thrives in such abundance lately) who delights in discovering things for himself, the author has therefore refrained from providing the usual list of "answers-in-the-back-of-the-book". The "answers" to all the exercises are to be found in the main body of the text and it is hoped that the student will form the habit of looking there for them.

It is customary at this point for authors to acknowledge their indebtedness and gratitude to their publishers, their old professors, their colleagues, their students, and their wives for the various appropriate forms of cooperation, inspiration and help provided. The present author is mindful of these debts, he is sincerely grateful for all the help received and he is happy to use this opportunity to express his thanks to all concerned. Having only one wife, Rachel, who helped with the art work, and only one publisher, Pergamon Press, the author can and does give his thanks to these individually. However, in listing our colleagues, which in a broad sense include both our teachers and our students, we should be aware of our indebtedness to all of that great company of thinking humans who from the days of the cave man down to the present have worked patiently and painstakingly to create, assemble and transmit to us that exquisite and powerful system of concepts which enables us to deal with the quantitative aspects of the universe. To all these we are grateful for the heritage they have given us.

O. M. KLOSE

Arcata, California

PART I

The Number Systems

CHAPTER 1.1

THE ORIGIN OF NUMERICAL THINKING

HUMAN BEINGS are sometimes reputed (by human beings) to have elevated themselves somewhat above the level of the other forms of animal life on this planet. If so, their chief basis for such a claim probably lies in the fact that they have taken upon themselves the task of trying to understand the nature of the physical universe—among other things. This endeavor and the various activities arising from it are collectively labeled as **science** (a word of Latin origin which may be roughly translated as "knowing").

In trying to understand the things of the physical universe one repeatedly encounters certain basic *patterns*. These may be patterns of *form* such as the shapes and paths of the astronomical bodies, they may be patterns of *arrangement* such as the symmetrical arrangement of the limbs of living creatures; or they may be patterns of *relation* such as the ordering in men's minds of sound, which we call music, the ordering in men's minds of form and color, which we call art, the ordering in men's minds of words, which we call poetry, or the ordering in men's minds of thought, which we call philosophy. Our study of these and all other such patterns is what we call **mathematics.**

Nearly all of mathematics involves the use of a concept called **number.** It is natural then that the study of mathematics should begin with, be centered upon, and be extensively and intimately concerned with numbers. The elementary study and use of numbers is called **arithmetic.**

Since a number is not a physical thing but an abstract concept (that is, a creation of thought) and one which has evolved thru the ages in the minds of thinking creatures, a good way to understand this concept is to retrace its development in the minds of primitive men who first used it for the practical purpose of keeping track of their sets of material acquisitions and personal belongings.

EXERCISES

This is the "do-it-yourself" department. The following items—and other similar ones which will appear later from time to time—are what are called "exercises" in most textbooks. We shall call them "exercises" too because an exercise is a voluntary exertion which is performed regularly for the purpose of establishing a useful reflex action and strengthening the muscles involved in this action, and we hope that you will use these "exercises" in exactly this way to establish the reflex mental actions associated with mathematical thinking and to strengthen your mathematical muscles. No athlete develops muscular coordination entirely by reading a book, listening to a lecture or attending a "workshop".† Similarly, the more you exercise and practice your mathematical skills, the better your game will be.

1. List ten sciences, professions, trades or regular human activities of any sort which do not make use of, and do not have need for, the application of any mathematical principles or concepts whatever.

2. List five different items among your personal possessions which you think might still exist and be in your possession even if you lived your life in a society with no awareness of mathematical concepts.

3. Discuss three examples of non-human forms of life which evidence either a (possibly unconscious) use of, or some apparent understanding of, mathematical principles.

4. Explain the meaning of the words (a) concept, (b) abstract. (In defining "abstract", define it as an adjective which might be used in modifying the noun "concept".)

5. (a) Without making any use of references or other help, write the best definition that you can compose, on the basis of your own present understanding, of the concept of "number". (b) Do the same for "set".

1.1.1. The Set Concept in Human Affairs

The concept of **set** is so useful in mathematics as to be considered of basic importance. Like other basic mathematical concepts, its meaning cannot be satisfactorily defined in terms of

† "Workshop": formerly (during the days of actual workshops) called a *conference*. Sometimes also called (irreverently, by undergraduates) a "bull-session".

other still more basic concepts (because there are no other concepts which are more basic) and, for this reason, no attempt is made to define it. Intuitively, its meaning can be conveyed quite easily and clearly merely by giving a few common synonyms for it such as *class*, *bunch*, *collection*, *lot*, *aggregate*, etc. Thus the class of people whose incomes exceed $50,000 per year is a set; a bunch of bananas is a set; a collection of paintings is a set; a lot of people in an auditorium is a set; the aggregate of all items in a lady's handbag is a set; etc.

The individual items or objects in a set are called the **elements** of the set. The elements of a set may have certain obviously recognizable common properties as, for example, in the case of a bunch of keys, or a collection of stamps, or a bouquet of roses. However, such similarity among elements is not necessary, as three apples and two oranges constitute just as good a set as would five apples—or seven oranges. For some sets it might be that the only common property possessed by all elements of a set is their common membership in the particular set being considered. (Consider, for example, the set of items in a lady's handbag.)

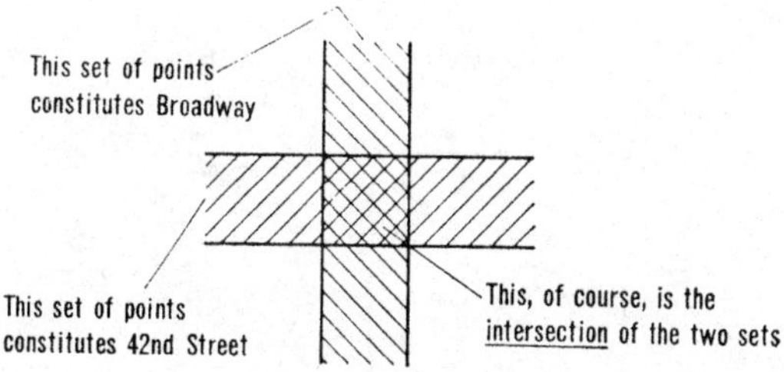

When two or more sets are under consideration, there are certain important other sets which may be constructed from them. One such set, called the **intersection** of the original sets, is the set consisting of all those elements each of which belongs to every one of the original sets. For example, if A is the set of all Republicans and B is the set of all Presbyterians, then the intersection of A and B is the set of all people each of whom is both a

Republican and a Presbyterian. As another example, a street intersection, as illustrated in the figure on p. 5, is also an intersection of sets in the mathematical sense.

A second important derived set is the **union** of the original sets, and this is the set consisting of all elements which belong to any one (or more) of the original sets. For example, the union of sets A and B as described in the preceding paragraph would be the set of all people who are *either* Republicans *or* Presbyterians *or both.* As another example, consider the following sets.

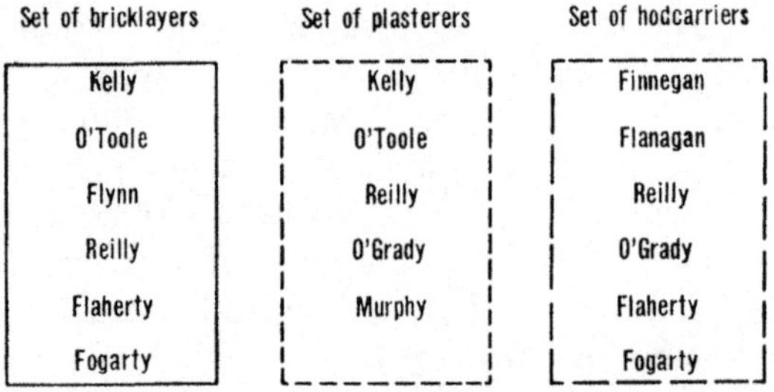

The union of these three sets is illustrated in the following diagrammatic list.

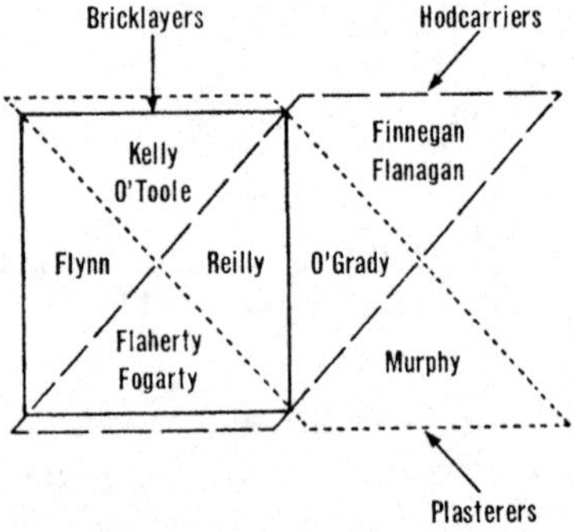

A set, all of whose elements are also contained in another set, is called a **subset**† of the latter set. For example, the set of all

† An alternative and slightly better definition would be: "A set P is a subset of a set Q if any element which is not in Q is also not in P." Do you see why?

Republicans is a subset of the set of all people, the set of all women is a subset of the set of all people, the set of all Eskimo women is a subset of the set of all women, the set of all plasterers is a subset of the set of all bricklayers, plasterers and hodcarriers, etc.

If, for some two sets A and B, B is a subset of A and A contains also at least one element which is not in B, then B is called a **proper subset** of A.

If a set S is a subset of a set T, then T is called a **superset** of S. If S is a proper subset of T, then T is a **proper superset** of S.

A set which contains no element is—naturally—called "the **empty set**", or "the **null set**". It might seem that if a set had no element then there would be no set to talk about. An example will show that this is not the case. Legendary reports from the Himalayan Mountains tell of creatures there which are called "yetis", or "abominable snowmen". Let us *define* the set Y (that is, let us *describe*, or *explain*, or *identify* the set Y) as the set of all yetis. Up to the present time it has not been proved that a single yeti exists, and it seems unlikely that any do exist outside the realm of tall stories. Thus the set Y may be, and probably is, the empty set; nevertheless it has been the subject of worldwide attention and discussion.

While the empty set is **unique** (which means that there is *just one* set of this kind) it may nevertheless be considered from various different viewpoints. Thus it may be regarded as the day's collection of eggs—for a day on which there was no egg to be collected. It may be considered as the set of all vacuum cleaners sold by a vacuum cleaner salesman who never actually sells any vacuum cleaner. It may be considered as the set of "little men who aren't there", the set of all hens' teeth, and so on. It turns out that the empty set, trivial as it may at first seem, has some rather important consequences.

If two sets have no common element, then their intersection set (being the set of all their common elements) is, of course, the empty set. In such a case we say that the two sets are **disjoint,** or that each of the sets is **disjoint, relative to the other.**

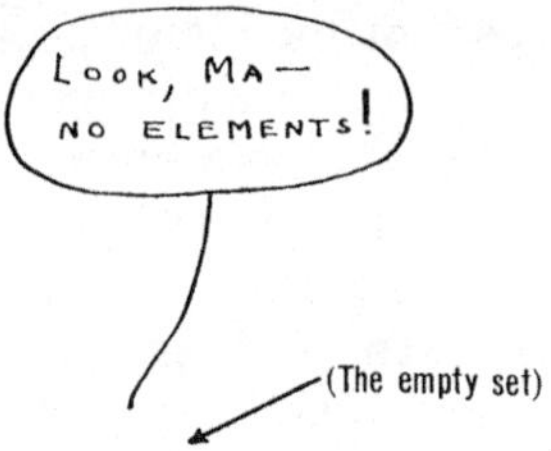

(The empty set)

In certain discussions it may be well understood that each element of every set discussed is also an element of some common superset E. In such situations the set E is called "the **universal set**", "the **universe of discourse**", or simply "the **universe**". For example, if we are discussing such sets as the set C of all Californians, the set B of all businessmen, the set D of all doctors, the set F of all Finns, etc., it is obvious that the universal set for our discussion is the set of all human beings.

If the union of two disjoint sets, say A and B, is a set E, then each of A and B is called "the **complement** of the other **relative to** E". If E is understood to be the universal set then it is customary just to say that A is "the **complement** of B", and vice versa.

1.1.1.1. *Set notation.*

In order to have a standard way of describing or identifying a particular set it has been agreed by most people who are concerned with the discussion of sets that a set should be represented by writing (or indicating in some manner) all the elements of the set within the pair of symbols { } called **braces.** Thus if set A consists of the five symbols 0, +, ✱, Δ and →, then we write

$$A = \{0, +, \bigstar, \Delta, \rightarrow\}.$$

The symbol " = " means "is", or "is the same set as", and is read as "equals". Similarly, if B is the set of three people George, Don and Buford, and C is the set of *all* the numbers 1, 2, 3 *and so on*, we write

$$B = \{\text{George, Don, Buford}\}$$
$$C = \{1, 2, 3, \ldots\}.$$

Note that in describing a set, as such, the order in which the elements are listed is not important. Thus, for instance, if $B =$ {George, Don, Buford}, then also

$B =$ {George, Buford, Don}
$B =$ {Don, George, Buford}
$B =$ {Don, Buford, George}
$B =$ {Buford, Don, George}
$B =$ {Buford, George, Don}.

As another example, we may now replace the pictorial or diagrammatic scheme used on page 6 in displaying the sets of bricklayers, plasterers and hodcarriers, and their union, by the use of standard set notation as follows.

Set of bricklayers = {Kelly, O'Toole, Flynn, Reilly, Flaherty, Fogarty}
Set of plasterers = {Kelly, O'Toole, Reilly, O'Grady, Murphy}
Set of hodcarriers = {Finnegan, Flanagan, Reilly, O'Grady, Flaherty, Fogarty}
Union of bricklayers, plasterers and hodcarriers = {Kelly, O'Toole, Flynn, Reilly, Flaherty, Fogarty, O'Grady, Murphy, Finnegan, Flanagan}

The union of the sets A and B is written as

$$A \cup B$$

and that of the sets A, B and C as

$$A \cup B \cup C, \text{ etc.}$$

Similarly, the intersections of set A and B and of sets A, B and C are written respectively as

$$A \cap B \text{ and } A \cap B \cap C.$$

The assertion that set A is a subset of set B may be written in either of the two symbolic forms

$$A \subseteq B, B \supseteq A.$$

The first of these two forms is read as "A **is included** in B" or as "A is a subset of B"; the second form is read as "B **includes** A" or as "B is a superset of A". Similarly,

$$A \subset B$$

means "A **is properly included** in B" or "A is a proper subset of B", while

$$B \supset A$$

means, equivalently, that "B **properly includes** A" or "B is a proper superset of A".

The complement of set A is written as A^c.

The empty set is usually written as ϕ.

EXERCISES. [Do them *yourself*!]

1. Master Patrick Joseph Kelly, 6 years old, asks: "What is a set?" Answer him.

2. (a) If the two sets of names, {Helen, Mary, Ann} and {Ruth, Rose, Mary}, are written merely as sets of girls' names, not necessarily representing any particular people, what is the union of the two sets?

(b) If the sets {Helen, Mary Kelly, Ann} and {Ruth, Rose, Mary Rafferty} represent actual people, what is the union of the two sets?

(c) Explain the essential difference involved in the two preceding cases.

3. What can you deduce about sets A and B if the intersection of A and B is set B?

4. What can you deduce about sets C and D if the union of C and D is set C?

5. What can you deduce about sets E and F if the complement of E relative to F is the set E?

6. The elements of set X are the present members of the United States Senate. Set Y is the set of all living yetis. What is the union of X and Y? What is the intersection of X and Y? (Assume that there are no living yetis other than those which you personally know and can vouch for.)

7. Compare $A \cup B$ and $B \cup A$.

8. Compare $A \cap B$ and $B \cap A$.

9. The notation $(A \cup B) \cup C$ means the result of first forming the union of A and B and then forming the union of this union and C. The notation $A \cup (B \cup C)$ means the union of A with the union of B and C. Compare each of these final resulting sets with the other and also with the union $A \cup B \cup C$.

10. Restate Exercise 9 for intersections rather than unions and then discuss the problem as restated.

11. Is the union of two sets always a set? Why? Is the intersection of two sets always a set? Why? Is the complement of a set always a set? Why?

12. Compare the sets $A \cap (B \cap C)$ and $(A \cap B) \cup (A \cap C)$.

13. What can you deduce about (a) $A \cap A$? (b) $A \cup A$? (c) $(A^c)^c$?

14. What can you deduce about (a) $A \cap \phi$? (b) $A \cup \phi$?

15. If E is the universal set, what can you deduce about (a) E^c? (b) ϕ^c?

16. How many different appearing ways are there of writing the set {0, +, ✱, Δ, →}?

17. Considering the words "gumbo", "homburg", "humble", "humbug", "lumber" and "thumb" merely as sets of letters, find (a) the intersection of these sets; (b) their union.

18. The empty set has been described as "the set of little men who aren't there", "the set of all hens' teeth", etc. Give three similar colloquial ways of describing it which have not been mentioned in the foregoing discussion.

19. List all the subsets (including both proper and improper subsets) of the set (a) $\{A,B\}$; (b) $\{A,B,C\}$; (c) $\{A,B,C,D\}$. (d) Deduce, if you can, the number of such subsets for the set $\{A,B,C,D,E\}$.

20. (a) Write the set A of all letters in the word "spontaneous".

(b) Write the set B of all letters in the word "combustion".

(c) Write the union of A and B (as a set of letters).

(d) Write the intersection of A and B (as a set of letters).

21. (a) List a dozen or more different items of any sort. Call the aggregate of these items set S.

(b) Identify, not by listing their elements but by describing some distinctive property of their elements, five subsets of the set described in (a). Label these five sets as A, B, C, D and E.

(c) What elements are in the intersection of B and D?

(d) What elements are in the union of A and B?

(e) What elements are in the complement of C relative to S?

22. For the sets of bricklayers, plasterers and hodcarriers as given on pages 6 and 9, list the members in each of the following sets:

(a) the union of bricklayers and plasterers;

(b) the union of bricklayers and hodcarriers;

(c) the union of plasterers and hodcarriers;

(d) the intersection of bricklayers and plasterers;

(e) the intersection of bricklayers and hodcarriers;

(f) the intersection of plasterers and hodcarriers;

(g) the intersections of bricklayers, plasterers and hodcarriers;

(h) the complement of the bricklayers.

23. Devise an exercise suitable for assigning to elementary school students involving application of the set concept. (Fit your exercise to any one of the first six grade levels, but specify which particular grade you have in mind.)

24. Give a set (by listing its elements) of at least six elements which have no similarity or common property beyond the fact that you are capable of conceiving of them mentally and that they are all in the set.

25. List the set of elements found in a small boy's pockets.

1.1.2. The Natural Number Concept

In dealing with sets of things even primitive men soon† find that some sets are more valuable, and hence also more desirable, than others; consequently a need is recognized for some reliable way of comparing one set with another. For instance, this set of clams

might be accepted in exchange for a fish, but this set of clams

might not. Hence the former set might be considered as "worth more" "more valuable", "better than" or "superior to" the latter. It would then become important in a clam-based economy to be able to compare one set of clams with another set of clams and to recognize unerringly the "more valuable" set.

If a progressive society of cave men were to concentrate on this problem hard enough and long enough in the manner of a "group learning situation", it is possible that after a moderate number of generations they might produce essentially a method such as the following for comparing sets of clams.

First arrange the sets in rows, side-by-side, like this.

† "soon": A relative term meaning, in this case, "in only a few ice ages".

Then **match** the elements of the sets as far as is possible in some manner, for example, by drawing lines from elements of one set to elements of the other set, like this.

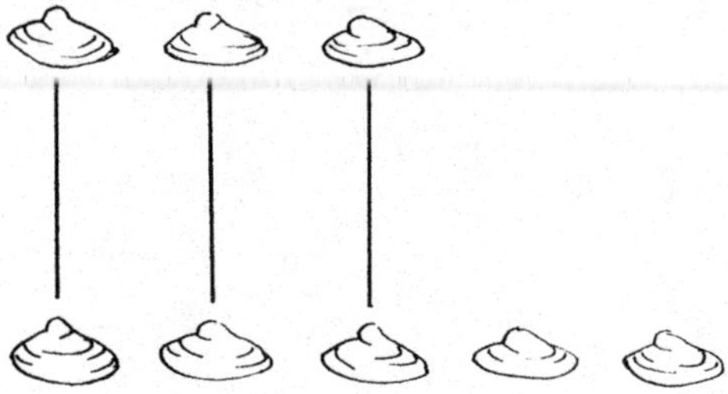

If this is done in such a way that *just one* element of one set is connected to *just one* element of the other set by each line drawn, and as many connecting lines are drawn in this way as is possible, **then one of the sets will be matched either to the other set or to some proper subset of it.** When two entire sets are matched in such a way, like this,

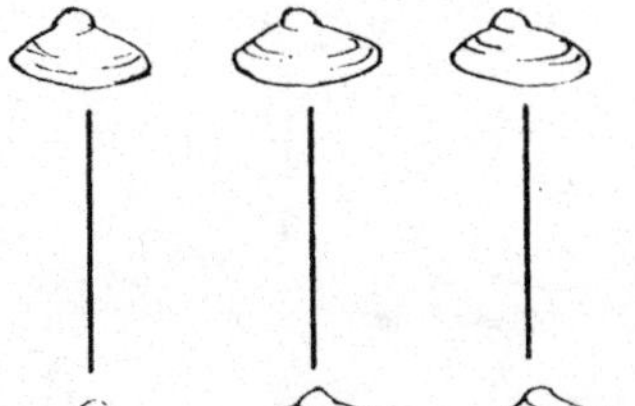

we say that they are in **one-to-one correspondence**; and the judgment of a cave-man society (or our own, for that matter) in such a case would probably be that the two sets had the same worth, or that one was about as good as the other. However, if the matching resulted in a one-to-one correspondence between one set (say set *A*) and a proper subset of the other set (say set *B*), so that after matching *some* of the elements of B with *all* of the elements of A there were still elements of B left over, like this,

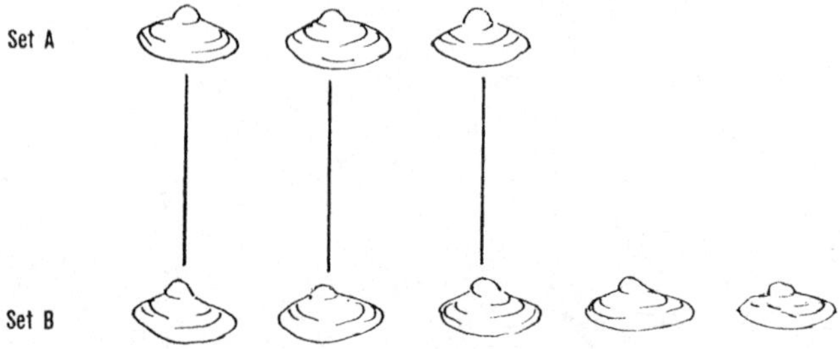

then the cave men would undoubtedly vote set B the more valuable and perhaps describe this situation by coining expressions such as "set B has **more** elements than set A", or "set B is **bigger** than set A".

An observant cave man might notice that this method would work as well for comparing sets of fish, thus,

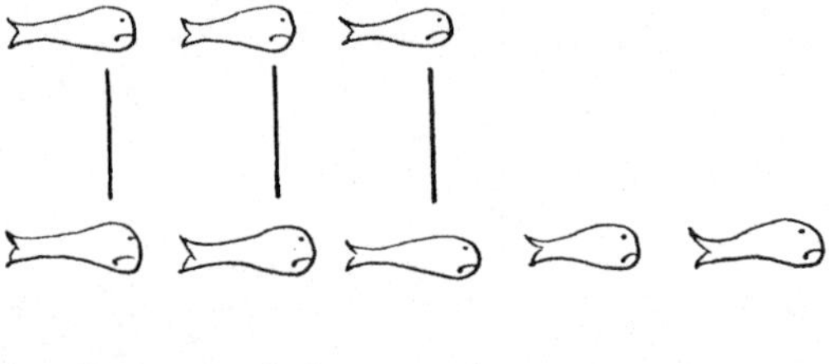

as for comparing sets of clams and, as a matter of fact, that it could be used for comparing sets of any sort of elements whatever. For instance, the military authorities of Bongo, before initiating military action against neighbouring Transbongo, might profit by making such a comparison of the armed forces of their own community and those of Transbongo, like this.

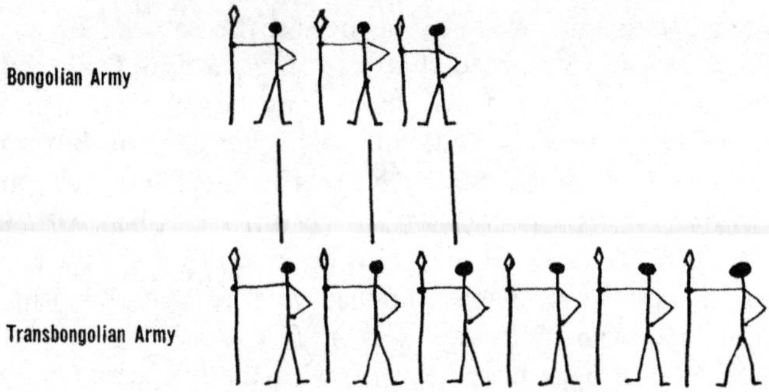

Such considerations are sometimes very practical.

In the course of making many such comparisons of sets an intelligent cave man might be expected to notice sooner or later that, while sets may be of various kinds, each set has a certain *distinguishing characteristic* quite apart from the nature of its

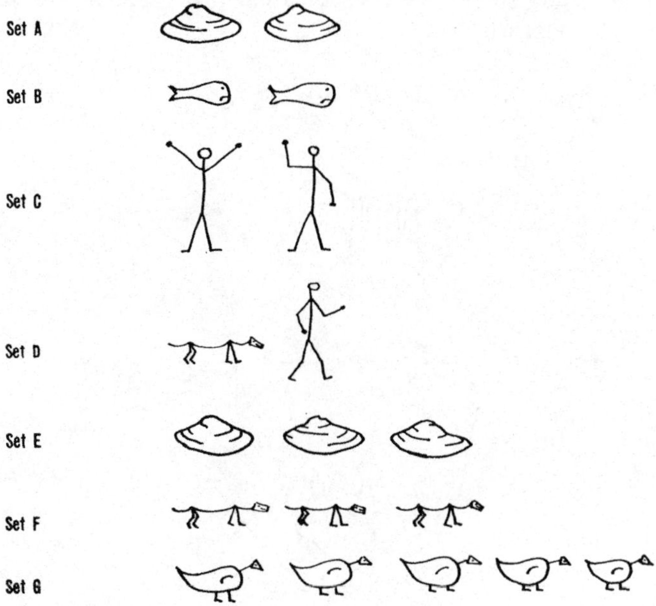

elements. For instance, in considering the sets *A*, *B*, *C*, *D*, *E*, *F* and *G*, it may be noted that sets *A*, *B*, *C* and *D* all share the distinction that any one of them may be put into one-to-one correspondence (or *matched*) with any other of them but not with any of sets *E*, *F* or *G*. Similarly, sets *E* and *F* may be put into one-to-one correspondence with each other but not with any of sets *A*, *B*, *C*, *D* or *G*. It may also be noted that set *G* cannot be put into one-to-one correspondence with any of the other sets. Saying this in another way, sets *A*, *B*, *C* and *D* form a **class of sets** (or a **set of sets**, if you prefer) with the property, or *characteristic*, that each set in the class can be matched with each other set in that class but not with any of the sets considered outside that class. Also, sets *E* and *F* form a class of sets with a characteristic of a similar (but distinct) sort, and set *G* is characterized by the fact that it cannot be matched with any of the other sets considered—altho it could, for instance, be matched with the set of fingers and thumb on a person's hand, or with the set of tentacles of a starfish.

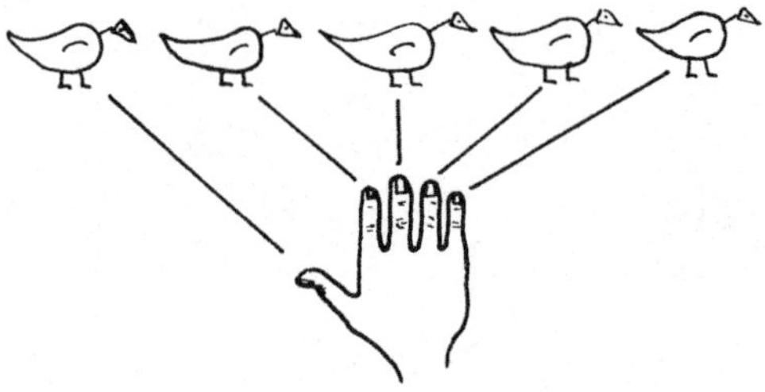

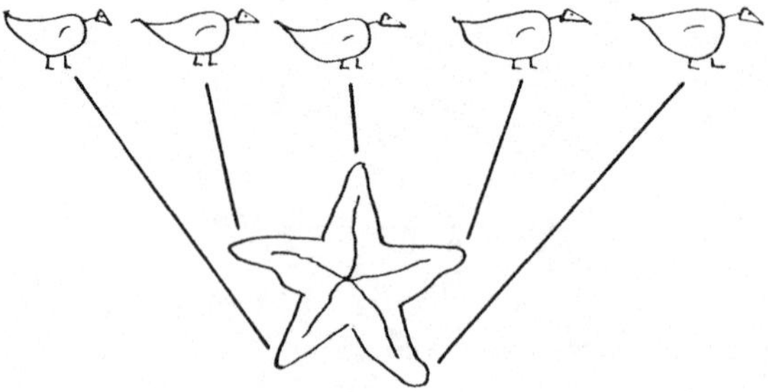

A further important thing to note is that this peculiar matching characteristic possessed by sets is not dependent at all on the nature of the elements in the sets. This being the case, we can disregard the actual nature of the elements in any set (as far as its matching characteristic is concerned) and represent any one of the sets *A*, *B*, *C* and *D*, for instance, by some simple standardized set of element symbols such as, possibly, these,

or perhaps these,

0 0

or even these,

· ·

Similarly, sets *E* and *F* could be represented by

0 0 0

or by

· · ·

and set *F* could be represented by

· · · · ·

Since it is simple and convenient to use one or another of the symbols 0,  and ·, and since by doing so we may more easily concentrate on the matter of the matching characteristic of a set without being distracted by considerations of whether the elements of the set are clams, fish, ducks, rabbits or men, we shall henceforth use such symbols freely in representing sets. Having thus streamlined our way of representing sets, we are now in a position to make a thoro and detailed survey of all the different kinds of set which have different matching characteristics.

First there are sets like this, ·
then there are sets like this, · ·
then sets like this, · · ·
then like this, · · · ·

then this, ...
and this, ...

and so on. The phrase "and so on" means that, like the sorceror's apprentice, we now realize that the thing we have started is not going to stop. There is obviously no limit to the different kinds of set, each with its peculiar associated matching characteristic. Since this makes somewhat difficult the attainment of our announced objective of investigating "all" such sets and matching characteristics, let us change the subject for the present by discussing **names** for some of the concepts we have been talking about.

First of all, what about a name for the general concept of "matching characteristic" of a set? People who like to play with words might like a contraction such as "matchstic"; others might prefer the abbreviation "m.c." or some other device. Fortunately this problem has been solved for the modern world by the development of language and, in particular, the English language, in which this concept is called a **natural number**, or sometimes simply a **number.** (Did you know this?) Having this much settled, our next concern is with the *particular* names for all the natural numbers associated with distinct kinds of sets. The following table shows both the **names** and the standard **symbols** which have been adopted in the English language for some of the various possible sets.

Kind of set	Matching characteristic, or "natural number"	
	Name of natural number	Symbol for natural number
.	One†	1
. .	Two‡	2
. . .	Three	3
. . . .	Four	4
. . . . .	Five	5
. . . . . .	Six	6
. . . .	Seven	7
. . . .	Eight	8
. . . .	Nine	9
. . . .	Ten	

† Pronounced "wun". (We don't know why either.)
‡ Here, of course, the "w" is silent. Pronounce this "Tue" as in Tuesday.

Again we seem to have encountered the predicament of the sorceror's apprentice. If we keep on going like this, we'll have to invent an unlimited set of names and symbols. We suspect this is not a practical undertaking; there must be an easier way.

Since a set like this,

• • • • • • • • • •

can be matched with the set of fingers and thumbs on both a person's hands,

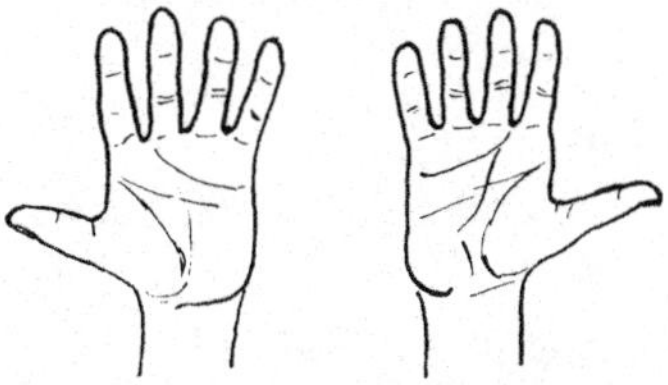

and any smaller set can be matched with some subset of this set of fingers and thumbs, we can represent any number from one to ten by displaying an appropriate set of fingers (and/or thumbs) and thus we seem to have things under fair control up to the number "ten". However, with ten names and nine symbols to remember already, it's time we concentrated on developing some labor-saving devices as an alternative to the invention of still other names and symbols.

1.1.2.1. *Positional notation, zero, digits and numerals*

Suppose we are faced with the necessity for finding a suitable name and symbol for the natural number associated with a set like this.

• •

Since the number ten is handy—because any set having this number as its matching characteristic can be put into one-to-one correspondence with the set of fingers and thumbs on both a person's hands—we might make use of it in the following way. First arrange the given set in as many subsets of ten elements each as possible and a smaller subset containing whatever elements are left over after the ten-element subsets have been set aside, like this.

• • • • • • • • • •   • • • • • • • • • •   • • •

Now having got two subsets of ten elements each and a subset of three extra elements, we could describe the associated natural number as made up to **two tens and a three** (which could conceivably be contracted into a condensed slang form such as "twenty-three") and we could represent it symbolically by writing

2 tens and 3

or simply

23

if we agree to understand that the 2, as written here, actually represents "two tens" while the 3 has just its ordinary meaning.

If the number of sets of ten elements each becomes, itself, ten or more, we could then assemble sets of **tens of ten elements** each (doing this as many times as possible), then assembling as many as possible leftover sets of ten elements each, and finally a set of less than ten leftover individual elements. We could then represent the associated natural number by writing down in order, from left to right, the number of sets of tens of ten elements, the number of sets of ten elements, and finally the number of individual elements. Thus the set

• • • • • • • • • •   • • • • • • • • • •   • • • • • • • • • •   • • • • •
• • • • • • • • • •   • • • • • • • • • •   • • • • • • • • • •   • • • • •
• • • • • • • • • •   • • • • • • • • • •   • • • • • • • • • •   • • • • •
• • • • • • • • • •   • • • • • • • • • •   • • • • • • • • • •   • • • • •
• • • • • • • • • •   • • • • • • • • • •   • • • • • • • • • •   • • • • •
• • • • • • • • • •   • • • • • • • • • •   • • • • • • • • • •   • • • • •
• • • • • • • • • •   • • • • • • • • • •   • • • • • • • • • •   • • • • •
• • • • • • • • • •   • • • • • • • • • •   • • • • • • • • • •   • • • •
• • • • • • • • • •   • • • • • • • • • •   • • • • • • • • • •   • • • •
• • • • • • • • • •   • • • • • • • • • •   • • • • • • • • • •   • • • •

could have its associated natural number written as

347,

meaning "three tens of tens, four tens, and seven". Similarly, the natural number symbol

25,426

would mean "two sets of tens of tens of tens of tens, five sets of tens of tens of tens, four sets of tens of tens, two sets of tens, and six". (It is customary, of course, to call ten tens a **hundred,** ten tens of tens a **thousand,** etc.) This method, in which the particular *position* of one of the symbols 1, 2, 3, etc., is used to show whether the symbol represents a number of individual elements, a number of tens of elements, a number of hundreds of elements, etc., is called the **positional method** of writing numbers, or **positional notation.** It can be extended to any length and seems to take care of the problem of representing any natural number symbolically without the introduction of any new number symbols beyond those we already have, namely 1, 2, 3, 4, 5, 6, 7, 8, 9.

There is just one objection to this system as we have considered it so far. How do we distinguish, for instance, between the natural number associated with sets like

. .

containing just two elements; the natural number associated with sets like

. . . .

containing just two tens of elements with no individual elements left over; the natural number associated with sets like

. . . .
. . . .
. . . .
. . . .
. . . .
. . . .
. . . .
. . . .
. . . .
. . . .

B

containing just two hundreds of elements but no extra tens of elements and no extra individual elements, etc.? Clearly we cannot write just "2" in each case without causing serious confusion. The "positional" principle seems to break down here since we do not have a number of individual elements in the second case, or either a number of tens or a number of individual elements in the third case to establish the *relative position* of the "2".

It is here that the empty set, of all things, salvages the situation. Remember "the set that isn't there"? First of all, since we have invented a name and a symbol for a number concept corresponding to each other possible set considered so far, let's do the same for the empty set by conjuring up a mental conception of an associated "number" which we shall call **zero** and represent by the symbol 0. Thus, when a set is empty, we shall say that "the number of its elements is zero".

Now the natural number of the set

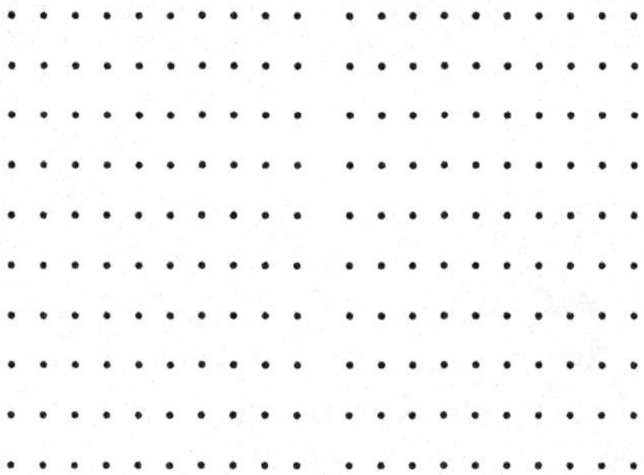

may be described as "two hundreds, zero tens, and zero" and written in exact accordance with our principle of positional notation as

200.

Similarly, the natural number of the set

is "two tens and zero" and is written as

20,

and the natural number of the set

• •

is, of course, written as

2.

The symbols 0, 1, 2, 3, 4, 5, 6, 7, 8, 9 are called **digits**. Any one of them or any combination of any or all of them used to represent a natural number is called a **numeral.** Any natural number whatever can be represented symbolically by some such numeral.

The system of digits just discussed, and their positional use to form numerals for representing natural numbers, is based on the use of the number ten in a key role. For this reason the resulting system of numerals is called a **"base-ten" system,** or a **"decimal"** or **"denary" system,** both of which latter terms, of course, mean the same thing as the former one. Numerical systems based on numbers other than ten could, of course, be designed in the same way. A society of number-conscious starfish would undoubtedly develop a system of digits and numerals based on the number five; a society of octopi would probably adopt a "base-eight" system; and a clam civilization with only the two parts of a clam shell to give it numerical inspiration would most naturally gravitate toward a "base-two" system and thus be psychologically well prepared for the use of high-speed digital computers which, as is well known, have a natural preference for arithmetic in this type of numerical system.

Exercises

1. Give three examples of one-to-one correspondence which are involved in the daily work of an elementary school teacher.

2. Why is there no "last" natural number?

3. Describe in detail a positional notation system for the natural numbers as it might have been developed by a number-conscious race of starfish.

4. Describe in detail a positional-notation system for the natural numbers as it might have been developed by a number-conscious race of octopi.

5. Describe in detail a positional-notation system for the natural numbers as it might have been developed by a number-conscious race of clams.

6. Invent a notation (other than the customary positional notation) for representing "ten tens", "ten tens of tens", "ten tens of tens of tens", etc. How could this notation be applied to represent ten itself? Could it be applied to represent one?

7. Interpreting the numbers 13, 51, 99, 100 and 289 as referring to base ten, convert each of them to the appropriate notation for (a) base five; (b) base eight; (c) base two.

8. Interpreting the numbers 13, 51, 100, 247 and 1214 as referring to base eight, convert each to the appropriate notation for (a) base two; (b) base seven; (c) base ten.

9. Interpreting the numbers 11, 101, 1100, 10101 and 1000000 as referring to base two, convert each to the appropriate notation for (a) base three; (b) base eight; (c) base ten.

10. We have discussed how two sets might be compared to determine which one is the "bigger". Discuss how two straight-line distances might similarly be compared. Are there any sort of number concepts which are associated with such straight-line distances in a manner similar to that in which the natural numbers are associated with sets of the type we have discussed?

1.1.2.2. *Some observations concerning sets and natural numbers*

Suppose we have a set S of some sort and suppose that, in order not to pin ourselves down to too particular a case, we call the natural number associated with this set n. Suppose also we have a set T consisting of just a single element (which is not in S), so that the natural number associated with T is the number *one*. Now suppose that we form the union of the disjoint sets S and T and call this union set S' (which is read as "S-prime", the mark " ′ " being called a **prime-mark**). Associated with the set S' there will be, of course, some natural number, say n'. We call this natural number the **successor** of the natural number n. Thus two is the successor of one, three is the successor of two, etc. Hence if, as in the figure, n is twelve, then

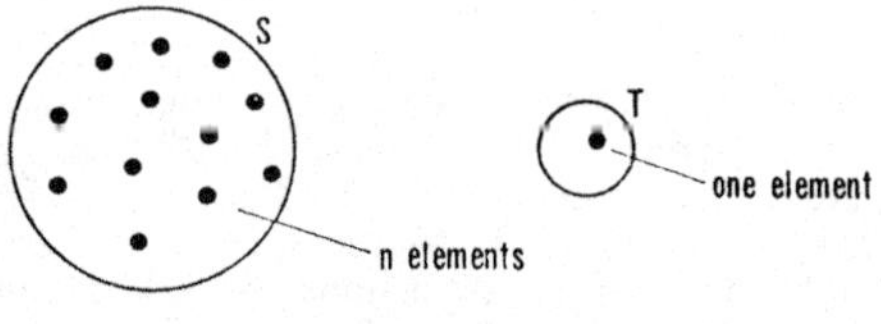

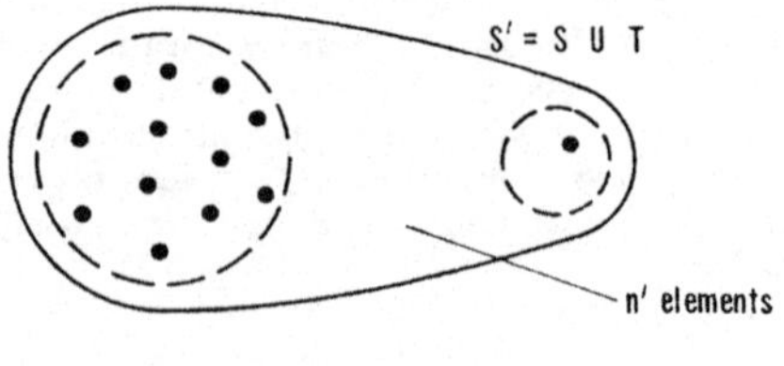

its successor n' is thirteen. If you were to interpret this as meaning that in general the successor of any particular natural number is just the "next higher natural number", you would be quite correct. Thus the idea of a successor is a quite simple one; and yet it turns out to be a rather important and useful one. If you think about it a little, you will probably agree to the truth of the following five statements.

(1) One is a natural number.
(2) For each natural number there is another natural number which is its successor.
(3) No two different natural numbers have the same successor.
(4) One is not the successor of any natural number. (This, of course, implies that we agree not to consider zero as included among the natural numbers.)
(5) If there is a set containing *some* (meaning "any or, possibly, all") of the natural numbers, which
    (a) contains the number one, and
    (b) for each natural number which it contains, contains also the successor of that number,

    then this set contains *all* of the natural numbers.

The first four facts stated are pretty simple and obvious. You may have to think about the fifth one a little. The interesting thing is that just about everything there is to say about numbers (and this includes all of arithmetic, most of elementary algebra and some other branches of mathematics) can be **derived,** or deduced as a consequence, from the above five simple statements. They are called **Peano's axioms.**

1.1.2.3. *Cardinal and ordinal numbers*

We have previously defined the natural number concept as a characteristic of a *whole* set—that is, a characteristic of *all* the elements of a set together. When the natural numbers are considered or used in this manner they are called **cardinal numbers.**

We now wish to point out a different use which is sometimes

made of the natural numbers. We shall illustrate this concept by referring to the set S shown in the following figure.

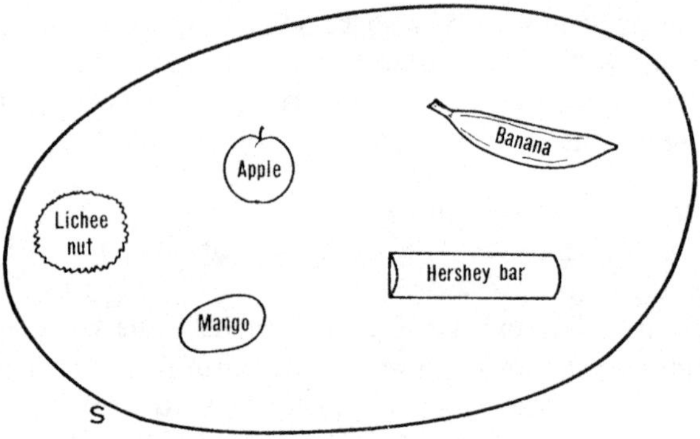

The elements of the set S might be arranged into a "nested" sequence of subsets in the following way. Suppose we begin by singling out some particular element—say the banana—for some reason (perhaps because we like bananas) and form a set A with the banana as its only element. Suppose we next direct our attention to some other element—say the Hershey Bar—and form a set B which includes A as a subset and also contains the Hershey Bar. Suppose we then form a set C which includes B as a subset and also contains some new element—say the apple. Similarly,

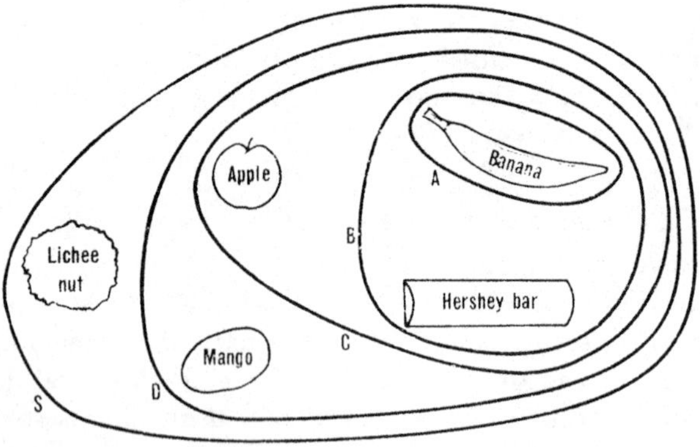

suppose we form a set D which includes C as a subset and which also contains some new element—say the mango. Finally, we note that the set S includes D as a subset and also contains the lichee nut.

The (cardinal) number associated with set A is 1, that with set B is 2, that with C is 3, with D is 4, and with S, 5. Notice that what is implied here is the following one-to-one correspondence between the set $\{A, B, C, D, S\}$ of nested sets and the set $\{1, 2, 3, 4, 5\}$ of cardinal numbers:

$$\begin{array}{ccc} A & \longleftrightarrow & 1 \\ B & \longleftrightarrow & 2 \\ C & \longleftrightarrow & 3 \\ D & \longleftrightarrow & 4 \\ S & \longleftrightarrow & 5. \end{array}$$

Recalling that, in constructing the sequence of sets A, B, C, D and S in that *order*, we began by fixing our attention on the banana element, then fixed it on the Hershey Bar element, then on the apple, then the mango, and finally the lichee nut, a different one-to-one correspondence is suggested, namely the following one between *the elements of* S and the set $\{1, 2, 3, 4, 5\}$:

Banana	⟷	1
Hershey Bar	⟷	2
Apple	⟷	3
Mango	⟷	4
Lichee nut	⟷	5.

Note that here the use of the numbers 1, 2, 3, 4 and 5 is not primarily to indicate the size of the sets A, B, C, D and S but to indicate the *order* in which the elements of S are considered. Numbers used in this way to indicate the *order* of selection (or of designation, or of consideration, etc.) are called **ordinal numbers.**

In dealing with ordinal numbers it is customary to replace the use of the noun forms "one", "two", "three", "four", etc., by the corresponding adjectival forms "first", "second", "third", "fourth", etc. Thus in the above example we have chosen the banana as the "first element", the Hershey Bar as the "second element", the apple as "third element", the mango as "fourth element" and the lichee nut as "fifth element".

EXERCISES

1. Determine the "successor of the successor" of each of the numbers 1, 2, 3, 5, 10, 15. Can you deduce a general statement concerning the successor of the successor of a given number? (*Do not use* any such terms as "add", "addition", "sum", "difference", etc., which have not yet been formally considered or defined.)

2. Do as in Exercise 1 for the "successor of the successor of the successor".

3. Do as in Exercise 1 for the "successor of the successor of the successor of the successor of the successor".

4. Find a short and convenient *terminology* for describing such concepts as "successor of the successor", "successor of the successor of the successor", etc.

5. In each of the following expressions identify the numbers which appear as either cardinal or ordinal. (a) George III; (b) mile 40; (c) six pounds for one dollar; (d) strontium 90; (e) Chanel No. 5; (f) 60 m.p.h.; (g) a 30 lb salmon; (h) 36-24-36; (i) the fourth republic; (j) the ten commandments; (k) Route 40.

6. Explain why we did not (and could not) define the "successor of a natural number n" by simply saying directly that it was the "next higher natural number".

1.1.3. The Counting Process

Primitive man's invention of number concepts not only served as a convenient means of *identifying* and *classifying* sets but also as a most useful means of *comparing* two sets.

We have shown that two sets (of clams, for instance) might be compared by lining them up and matching the elements of one set with those of the other set as far as possible, like this.

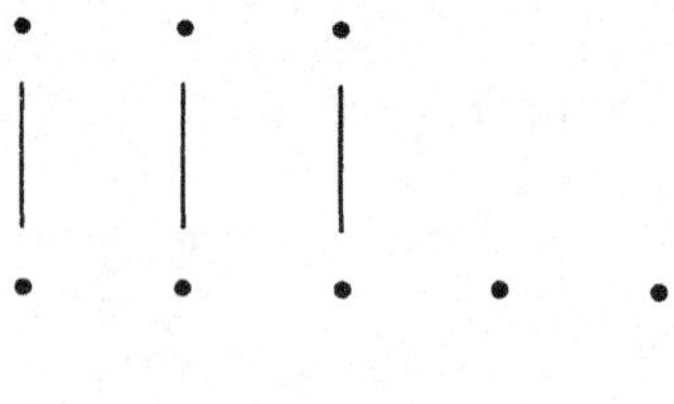

A method which has proved to be more efficient in general, and by means of which all such comparing processes can be standardized, can be developed by matching all sets with certain subsets of the set of natural numbers.

According to Peano's fifth axiom we could obtain a list of all the natural numbers by writing the number one, then writing the successor of one (which, of course, is two), then writing the successor of two (which is three), and so on. This order of listing the natural numbers,

$$1\ 2\ 3\ 4\ 5\ 6\ 7\ 8\ 9\ 10\ 11\ \ldots,$$

might be called their **natural order**—and we shall call it that. Let us also use the name **leading subset** to describe the following special type of subset which may be selected from the set of natural numbers:

First leading subset:	$\{1\}$
Second leading subset:	$\{1, 2\}$
Third leading subset:	$\{1, 2, 3\}$
Fourth leading subset:	$\{1, 2, 3, 4\}$
Fifth leading subset:	$\{1, 2, 3, 4, 5\}$, etc.

If now we arrange a one-to-one correspondence, or *matching*, between the elements of a specified set and the elements of some leading subset of the natural numbers, then **the number of elements in the specified set may be determined as the last natural number in the leading subset used.**

Examples

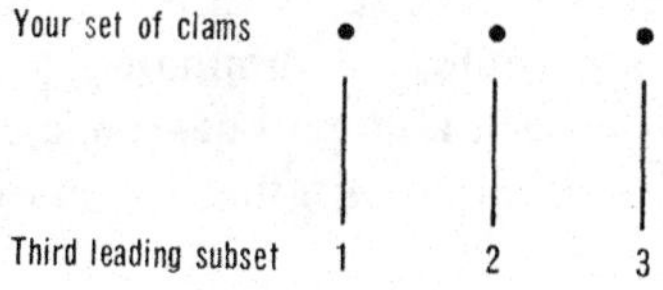

Conclusion: The number of clams in your set is *three*.

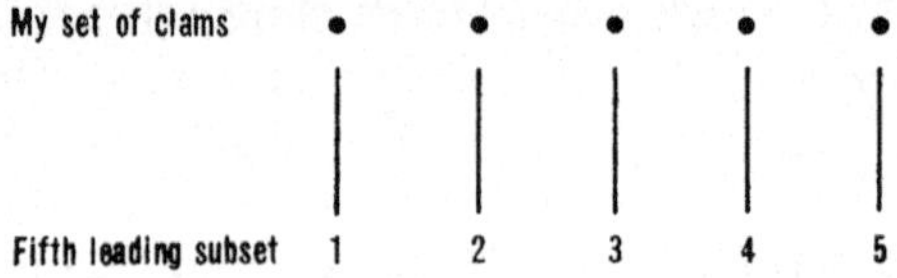

Conclusion: The number of clams in my set is *five.*

This process of making a more-or-less direct matching of a specified set with a leading subset of the natural numbers is called **counting** the specified set. We say "more-or-less direct" because some people do not do this in a purely direct manner but with the aid of some *intermediate* set of elements such as a set of fingers, or the numerical representation built into a computing machine, etc.

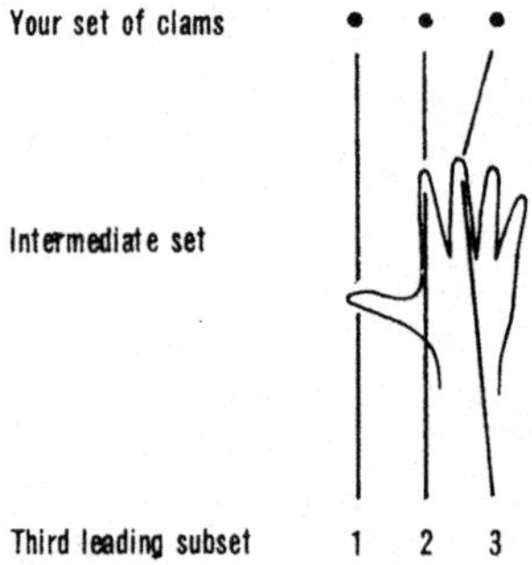

Any matching of a specified set with a leading subset of the natural numbers, whether direct or indirect (and involving the use of some standard intermediate reference set), is recognized as **counting**; and since whenever an indirect correspondence is used the intermediate set could be eliminated to produce a direct correspondence, **it is such a direct one-to-one correspondence that we shall keep in mind as constituting the essential nature of the counting process.**

To return to the problem of comparing your set of clams with my set of clams, we note that you have three clams, I have five clams, and five is the successor of the successor of three. Whenever one number is, in this manner, the end link in a chain of successors emanating from another number, we might agree with our cave-man ancestors to call the former number **greater** than the latter, and to say accordingly that that one of two sets which has the greater number of elements is **bigger,** or has **more** elements in it, than the other one (and is, therefore, usually more valuable and more desirable). Thus my set of five clams is *bigger* (and, therefore, better) than your set of three clams. The advantage of counting in such a case is that if both sets have been counted, then the conclusion (as to which set is bigger) can be reached easily without the necessity of lining up both sets of clams side-by-side and actual physical matching.

EXERCISES

1. Sets A and B are defined as follows.

A = {pencil, notebook, pen, letter opener, blotter}
B = {rifle, bayonet, canteen, gas mask, hand grenade, mess kit}

Making formal use of the definition given for counting: (a) count the number of elements in set A; (b) count the number of elements in set B; (c) count the number of elements in $A \cup B$.

2. Sets A and B are defined as follows.

A = {pencil, notebook, pen, typewriter, telephone}
B = {rifle, typewriter, canteen, gas mask, telephone, mess kit}

Making formal use of the definition given for counting: (a) count the number of elements in set A; (b) count the number of elements in set B; (c) count the number of elements in $A \cup B$.

3. Explain the essential difference between Exercises 1 and 2.

4. (a) Count the number of elements in the set {Ford, General Motors, Chrysler} in six different ways. (b) In how many different ways can the number of elements in the set {Ford, General Motors, Chrysler, Rambler} be counted?

5. Give three *new* examples of counting by means of an intermediate reference set.

CHAPTER 1.2

THE NATURAL OPERATIONS WITH THE NATURAL NUMBERS

AS HAS just previously been pointed out, the process of counting is useful for comparing two sets to determine whether one of them has the same number of elements or a greater number of elements in it than the other. As a matter of fact, of course, the invention of the counting process was motivated by the necessity of having some practical method of making such comparisons.

Our description of the result of comparing two sets (either directly or thru counting) can be simplified by using a generalization of the successor concept based on the use of ordinal numbers. Calling the "successor of the successor" of a number its "second successor", the "successor of the successor of the successor" the "third successor", etc., and referring to all such as "generalized successors", we can restate the conclusion of page 31 by saying that one number is **greater** (or **bigger**) than a second number if the first one is any generalized successor of the second. For example, eight is greater than three because eight is the fifth successor of three. (Note that in using the "generalized successor" terminology, the successor of a number would be called its "first successor".)

A new problem involving counting arises when the union of two disjoint sets is formed as, for instance, when Jo and Flo both put their clams in the same bucket. If Jo has a set of three clams and Flo has a set of five clams, the question might be raised concerning how many clams they have together. The process of combining the numbers three and five to obtain the number of clams in the union set in this case is an example of what we call **adding** (or "the operation of addition").

1.2.1. Addition Considered as Extended Counting

The fact that Jo has three clams and Flo has five means that a one-to-one correspondence can be arranged between Jo's set of clams and the third leading subset of the natural numbers and a similar correspondence can be arranged between Flo's set and the fifth leading subset. In applying the counting process directly

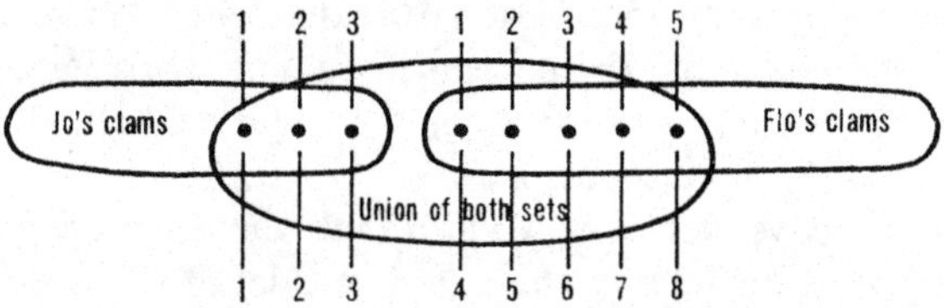

to the union set, we find that it can be put into one-to-one correspondence with the eighth leading subset of the natural numbers and hence that the number of clams in the union is eight. We may also note at the same time that what is essentially involved here (in arriving at the eight) is the substitution of the subset {4, 5, 6, 7, 8} in place of the leading subset {1, 2, 3, 4, 5} and the continuation of the counting process thru the subset {4, 5, 6, 7, 8} for Flo's clams after completing the count thru the leading subset {1, 2, 3} for Jo's clams. Thus the addition of three and five to get eight is just a matter of counting three clams (or three of any sort of element, for that matter) in the ordinary way and then *extending*, or *continuing*, the count thru the other five clams using the next (or *succeeding*) five natural numbers selected in their natural order beyond three.

In the language of generalized successors, we may describe the above process of obtaining the count of eight for the union set as the establishing of a one-to-one correspondence between the leading subset {1, 2, 3, 4, 5} and the set consisting of the first, second, third, fourth and fifth successors of three. From the conclusion, that **the number of elements in the union of a set of three elements and a disjoint set of five elements is the fifth successor of three** (and hence eight), the general rule of procedure for solving all problems of a similar nature can be easily deduced.

While this is all that is involved in addition, and any problem in addition (given the number of elements in each of two disjoint sets, to find the number of elements in their union) can be solved straightforwardly in this manner, it must be admitted that doing so would soon become tedious and inspire the observation: "There must be a better way".

It is an important principle of nature that whenever we reach the stage of awareness that there needs to be, and, therefore, must be, a better way of doing something, the required better way soon materializes. So it was with addition.

Observant cave men soon noticed that whenever five was added to three (that is, whenever the union of a set of three elements of any sort and a disjoint set of five elements of any sort was formed), the procedure was invariably the same and the resulting number of elements in the union was likewise invariably eight. Naturally, as soon as this fact became apparent, it was a simple matter to chalk this result up on the wall of the cave for handy future reference, thereby eliminating the necessity for ever having to add five to three by the long hard method again.

This labor-saving discovery so fascinated the academic members of the community that one primitive research-minded number theorist named Ho made a lengthy investigation to determine the nature of the results (called **sums**) obtainable by adding any two numbers and published a dissertation (on the wall of his cave) embodying a partial listing of such results. (See opposite page.)

This tabulation of addition results is sometimes called an **addition table.** The symbol " + " in the upper left corner represents the operation of addition and is spoken as "plus". To use this table to find the result of adding five clams to three clams we first locate the number three (3) in the left-hand column and the

+	1	2	3	4	5	6	7	8	9
1	2	3	4	5	6	7	8	9	10
2	3	4	5	6	7	8	9	10	11
3	4	5	6	7	8	9	10	11	12
4	5	6	7	8	9	10	11	12	13
5	6	7	8	9	10	11	12	13	14
6	7	8	9	10	11	12	13	14	15
7	8	9	10	11	12	13	14	15	16
8	9	10	11	12	13	14	15	16	17
9	10	11	12	13	14	15	16	17	18

number five (5) in the top row. Directly to the right of the 3 and directly under the 5 is found the number eight (8) which indicates that the result of adding five clams to three clams is eight clams. The essential numerical aspect of this relation is stated as "three plus five is eight" and this may be written in the briefer symbolic form

$$3 + 5 = 8.$$

The symbol " = " (which is read as "equals") is called the **symbol of equality,** or the **equals sign,** and means "is", or (in the present context) "is the same number as". Numbers which are combined in addition to form a sum are called **summands**; they are also called **addends**, and **terms**.

Examples of use of table:

$$2 + 4 = 6$$
$$2 + 7 = 9$$
$$3 + 6 = 9$$
$$4 + 5 = 9$$
$$5 + 4 = 9, \text{ etc.}$$

EXERCISES

1. (a) Find the first, second, sixth and thirteenth successors of five. (b) Find the fifth successors of one, two, six and thirteen. (c) What conclusion might you draw from observation of the results in parts (a) and (b)?

2. From the addition table on page 35 find the values of the following sums: (a) $3 + 4$; (b) $4 + 3$; (c) $3 + 8$; (d) $8 + 3$; (e) $4 + 9$; (f) $9 + 4$. (g) What conclusion might you draw from observation of the preceding results?

3. Evaluate the following sums: (a) $3 + 3$; (b) $6 + 3$; (c) $9 + 3$; (d) $4 + 4$; (e) $8 + 4$. (f) What conclusion might you draw?

4. Evaluate the following sums: (a) $2 + 3$; (b) $5 + 4$; (c) $3 + 4$; (d) $2 + 7$. (e) What conclusion might you draw?

5. Is the sum of every two natural numbers also a natural number? Why?

6. What meaning would you attach to the indicated sum $4 + 0$? To $0 + 4$? Explain.

7. On page 34 it is asserted that "the general rule of procedure for solving all problems (in the addition of two natural numbers) can be easily deduced". Deduce and state this rule for two arbitrary natural numbers. (The use of the word "arbitrary" here means that you are not to confine your statement to any two particular natural numbers such as, for instance, four and seven, but must state it in such a way as to be equally applicable to any and all pairs of natural numbers.)

8. Can you explain *why* the sums $2 + 3$ and $3 + 2$ have the same value?

9. Explain carefully what meaning (or meanings) you might attach to the expression $3 + 4 + 6$.

10. Construct the essential part of the addition table for the natural numbers in the octal (base eight) system. (By "essential part" we mean that part involving all cases in which both summands are one-digit numbers.)

11. Construct the essential part of the addition table for the natural numbers in the quinary (base five) system.

12. Construct the essential part of the addition table for the natural numbers in the binary (base two) system.

13. Perform the indicated additions, interpreting all given numbers as being in the octal (base eight) notation: (a) $5 + 7$; (b) $27 + 72$; (c) $306 + 624$; (d) $4357 + 526$.

14. Perform the indicated additions, interpreting all given numbers as being in the binary (base two) notation: (a) $11 + 10$; (b) $101 + 110$; (c) $11{,}011 + 101{,}101$.

15. Using the binary system, what number must the symbol n represent if $n + 10{,}111 = 111{,}110$?

16. Using the octal system, what number must the symbol m represent if $2765 + m = 42{,}373$?

1.2.1.1. *Addition of zero and a natural number*

One rather important bit of information not included in the addition table of page 36 but discovered some time later involved the number zero. If, for instance, we form the union of a set of three elements and the empty set (containing zero elements) and count the number of elements in this union, we find that it is also three, and hence that

$$3 + 0 = 3.$$

Likewise,

$$\begin{aligned} 0 + 3 &= 3, \\ 7 + 0 &= 7, \\ 0 + 7 &= 7, \text{ etc.} \end{aligned}$$

We thus observe that zero has the unusual and interesting property that when zero is combined in addition with any number, the resulting sum is just that number itself.

1.2.1.2. *Extended addition*

Situations frequently arise in which it is necessary to add not just two numbers, but many of them. These situations, of course, involve the formation of unions of many sets, each pair of which is disjoint. For instance, if Jo dug five clams, Mo three, Bo four and Thoreau seven, they might wish to know how many clams they had altogether. This problem could be solved by adding three to five to get eight, then adding four to eight to get twelve, and finally adding seven to twelve to get nineteen. The process could be shown in symbols as

$$\begin{aligned} 5 + 3 + 4 + 7 &= (5 + 3) + 4 + 7 \\ &= 8 + 4 + 7 \\ &= (8 + 4) + 7 \\ &= 12 + 7 \\ &= 19, \end{aligned}$$

the parentheses ( ) being used to indicate clearly which two numbers are actually to be added next.†

† Whenever, in an expression involving more than one operation, parentheses are placed around the indicated result of combining some two numbers by means of a specified operation, this is to be interpreted as meaning that the operation indicated *within* the parentheses is to be performed *before* the performance of any operations not included within parentheses. Brackets,

1.2.2. Multiplication Considered as a Special Case of Extended Addition

An especially simple problem of this sort sometimes occurs in which all the numbers to be added are the same. Thus if Jo, Mo, Bo and Thoreau each dug five clams, then the total number of clams dug by all of them would be obtained as

$$5 + 5 + 5 + 5$$

which, of course, could be evaluated thus:

$$\begin{aligned} 5 + 5 + 5 + 5 &= (5 + 5) + 5 + 5 \\ &= 10 + 5 + 5 \\ &= (10 + 5) + 5 \\ &= 15 + 5 \\ &= 20. \end{aligned}$$

This sum of *four fives* is an example of what is called a **multiple** of five. Other multiples of five are two fives (which, of course, is ten), three fives (or fifteen), five fives (or twenty-five), twelve fives, etc. Multiples of numbers turn out to be both interesting and useful and, for that reason, important. The process of identifying the value of a multiple which (as we have seen in the example of determining that four fives is twenty) is an extended application of addition has come to be called **multiplication.** Thus we say that the process of finding the value of four fives is "the multiplication of five by four" and the resulting sum (twenty) is called "the **product** of five multiplied by four". This process and the

---

[ ], and braces, { }, may also be used with the same significance. All three kinds of symbol—parentheses, brackets and braces—when used for this purpose are called **grouping symbols.**

EXAMPLE:

$(2 + 3) + (4 + 5) = 5 + 9 = 14$
$2 + [3 + (4 + 5)] = 2 + [3 + 9] = 2 + 12 = 14$
$[2 + (3 + 4)] + 5 = [2 + 7] + 5 = 9 + 5 = 14$
$2 + \{[3 + (4 + 5)] + 6\} = 2 + \{[3 + 9] + 6\} = 2 + \{12 + 6\} = 2 + 18 = 20.$

Grouping symbols have the same general significance when used with operations (other than addition) which will be discussed later.

relation of the numbers four and five to the resulting product twenty are described in various other ways also, such as "five taken four times (and added) is twenty" or, in a peculiar brand of pidgin English, which has become accepted thru common usage, "four times five is twenty".

A shorthand notation for representing the product "four times five" is

$$4 \cdot 5,$$

the dot indicating that these two numbers are to be used in multiplication (that is, in repeated addition), the numeral 5 indicating the specific number which is to be used in addition, and the numeral 4 indicating the number of times which the number five is to be so used. Notice that, in order to avoid confusion, the multiplication dot is written perceptibly higher than the level at which a period would be written.† In elementary discussions of mathematics the cross symbol " $\times$ " is frequently used with the same meaning as the dot we have used. Thus we may write either

$$4 \cdot 5 = 5 + 5 + 5 + 5 = 20$$

or

$$4 \times 5 = 5 + 5 + 5 + 5 = 20.$$

This product is called the **fourth multiple of five.** The numbers four and five as used here are called the **multiplier** and the **multiplicand** respectively, and both are called **factors** of the resulting product.

1.2.2.1. *Products involving one as a factor*

Since the determination of the product $3 \cdot 5$ is essentially the counting of elements in the union of three disjoint sets of five elements each,

† The placing of the multiplication dot symbol at a level higher than that of a period is not universal. In England, for instance, the multiplication dot is placed at the same level as a period would be and decimal points are placed at the higher level where, in the United States, the multiplication dot would appear.

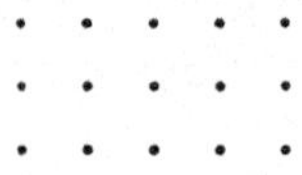

and the determination of the product 2 · 5 is the counting of the elements in the union of two disjoint sets of five elements each,

it is reasonable to interpret the determination of the product 1 · 5 as the counting of the elements in one set of five elements,

which leads us to recognize that 1 · 5 = 5.

It follows, similarly, that the determination of the product 5 · 1 is the counting of the elements in the union of five disjoint sets of one element each. Clearly this is five. Hence 5 · 1 = 5.

Any product involving the number one as a factor would be interpreted in a similar manner.

1.2.2.2. *Products involving zero as a factor*

Interpreting the products 0 · 5 and 5 · 0 in accordance with our definition of multiplication as extended addition (and hence as the counting of the elements in certain sets), the value of 0 · 5 is the counted number of elements in the union of zero sets (that is, no set at all) of five elements each. Such a union would contain zero elements (that is, no element at all) and hence 0 · 5 = 0. Similarly, the value of 5 · 0 is the counted number of elements in the union of five sets of zero elements each. This union would also contain zero elements so that 5 · 0 = 0.

A similar interpretation would be made for any product involving the number zero as a factor.

Just as the procedure for adding two numbers always leads to the same sum for the same two numbers and these sums, once determined, can be tabulated for convenient future reference, the same is true for the products obtained in multiplying pairs of numbers. The following table which gives the essential facts for such products (in base-ten notation) is called a "table of multiplication facts" or, more simply, a **"multiplication table"**. Reference to this table has provided a convenient short cut in multiplication problems since its discovery in cave-man days to the present time.

·	0	1	2	3	4	5	6	7	8	9
0	0	0	0	0	0	0	0	0	0	0
1	0	1	2	3	4	5	6	7	8	9
2	0	2	4	6	8	10	12	14	16	18
3	0	3	6	9	12	15	18	21	24	27
4	0	4	8	12	16	20	24	28	32	36
5	0	5	10	15	20	25	30	35	40	45
6	0	6	12	18	24	30	36	42	48	54
7	0	7	14	21	28	35	42	49	56	63
8	0	8	16	24	32	40	48	56	64	72
9	0	9	18	27	36	45	54	63	72	81

EXAMPLES

1. We have formally defined the meaning of the extended sum $2 + 3 + 4$ by writing $2 + 3 + 4 = (2 + 3) + 4$, where the use of the parentheses indicates that the 2 and the 3 are to be added first and the 4 is then to be added to the sum resulting from the first addition. Compare the value of $(2 + 3) + 4$ with that of $2 + (3 + 4)$. Explain *why* these two values are related as they are. (In your explanation be careful that you *do not merely assert* that something is true. Instead, be sure to give a genuine *explanation* of the *reason why* it is true.)

2. Compare the conceptual nature of the product $2 \cdot 3$ with that of the product $3 \cdot 2$. By comparing the values of these two products and other similar pairs such as $3 \cdot 4$ and $4 \cdot 3$, etc., can you state a general conjecture concerning the products $m \cdot n$ and $n \cdot m$ (where the symbols m and n are understood as representing any pair of natural numbers)?

3. All numbers in this problem are to be considered as given in octal notation. (a) What number must the symbol b represent if $2 \cdot b = 112$? (b) What number must the symbol c represent if $12 \cdot c = 346$? (c) Is there a natural number, say x, such that $3 \cdot x = 42$? If so, what is it? (d) Draw a representation of a set associated with the product $4 \cdot 12$.

4. All numbers in this problem are to be considered as given in binary notation. (a) What number must the symbol b represent if $11 \cdot b = 1111$? (b) Is there a natural number, say x, such that $11 \cdot x = 1101$? (c) Draw a representation of a set associated with the product $100 \cdot 1010$.

5. Construct the essential multiplication table for the binary notation.

6. Construct the essential multiplication table for the octal notation.

7. Construct the essential multiplication table for the base-six notation.

8. Explain carefully what meaning (or meanings) might be associated with the expression $4 \cdot 5 \cdot 7$.

9. Evaluate the expressions: (a) $2 \cdot (4 + 3)$; (b) $(2 \cdot 4) + (2 \cdot 3)$. [Note: The use of the parentheses in (a) indicates that it is the result of adding 3 to 4 that is to be multiplied by 2. In (b) the parentheses indicate that the result of multiplying 3 by 2 is to be added to the result of multiplying 4 by 2.] (c) What general conclusion can you draw from observing the final results in (a) and (b)?

1.2.2.3. *Extended multiplication*

Just as the process of addition can be used repeatedly with many numbers (as, for instance, in $5 + 3 + 4 + 7$), in a similar manner the operation of multiplication can also be used repeatedly with many numbers. As an example, the extended product $5 \cdot 3 \cdot 4 \cdot 7$ could be interpreted like this,

$$\begin{aligned} 5 \cdot 3 \cdot 4 \cdot 7 &= 5 \cdot 3 \cdot (4 \cdot 7) \\ &= 5 \cdot 3 \cdot 28 \\ &= 5 \cdot (3 \cdot 28) \\ &= 5 \cdot 84 \\ &= 420, \end{aligned}$$

the parentheses being used to indicate which two numbers are to be multiplied next. That is, we take seven four times (and add) to get twenty-eight, we then take twenty-eight three times to get eighty-four, and finally we take eighty-four five times to get four hundred twenty.

1.2.3. Exponentiation Considered as a Special Case of Extended Multiplication

It sometimes happens, and rather frequently, that we have occasion to consider cases of extended products in which all the factors are the same number. For instance, we might need to consider the product $5 \cdot 5 \cdot 5 \cdot 5$ whose value, of course, would be found thus:

$$\begin{aligned} 5 \cdot 5 \cdot 5 \cdot 5 &= 5 \cdot 5 \cdot (5 \cdot 5) \\ &= 5 \cdot 5 \cdot 25 \\ &= 5 \cdot (5 \cdot 25) \\ &= 5 \cdot 125 \\ &= 625. \end{aligned}$$

Special extended products of this sort are found to be just as important as the special extended sums which we call products. Consequently, they are equally deserving of special terminology, special notation and the other special treatments accorded important ideas in mathematics. Such a special extended product is called a **power.** In particular the product $5 \cdot 5 \cdot 5 \cdot 5$ is called the **fourth power of five.** It is represented by the shorthand notation 5^4 in which the numeral 5 indicates the number which occurs repeatedly as a factor and the smaller numeral ^{4} indicates the number of times which the number five occurs as a factor. The number five, as it is used in this example, is called the **base** of the power and the number four is called the **exponent** of the power. The process of forming a power is called **exponentiation.**

1.2.4. Binary Operations

Any process which is used to combine two numbers to obtain a third number is called a **binary operation** (the word "binary" meaning "two-part"). Since each of the three operations of addition, multiplication and exponentiation is of this type, these three processes are, of course, binary operations.

Since the need for, and the use of, the binary operations of addition, multiplication and exponentiation arise quite naturally in mankind's dealings with the natural numbers and in keeping the accounts for the various sets of objects in which mankind is interested, it seems natural to refer to these three operations as the **natural operations,** and we shall do so.

EXERCISES

1. (a) Explain the *meaning* to be associated with each of the expressions $2 \cdot (3 \cdot 4)$ and $(2 \cdot 3) \cdot 4$. (b) Find the *numerical value* for each of the expressions given in (a). (c) What conclusion might you draw from your observation of the results in (b)?

2. (a) Find the value of each of the following: 4^2 10^2, 10^3, 3^3, 3^4, 2^4, 2^{10}. (b) State a general conjecture concerning the values of m^n and n^m, where the symbols m and n are understood to represent any two natural numbers.

3. (a) What number must p represent if $p^5 = 243$? (b) What number must q represent if $3^q = 243$? (c) What number must u represent if $4^u = 4096$? (d) What number must v represent if $v^3 = 343$?

4. (a) Is there a natural number r such that $r^2 = 10$? Is there any such number at all? (b) Is there a natural number s such that $4^s = 17$? Is there any such number at all?

5. What meaning might you associate with 5^0?

6. Make an exponentiation table with both base and exponent taking each value in the set $\{1, 2, 3, 4, 5, 6, 7, 8, 9\}$. What important differences do you notice between this table and the tables for addition and multiplication (on pages 35 and 41 respectively)?

7. *Describe* two binary operations different from those already mentioned.

8. Explain how the use of certain powers plays a part of basic importance in our dealings with numbers.

CHAPTER 1.3

THE NATURAL LAWS FOR THE NATURAL OPERATIONS

IN USING the natural operations on the natural numbers (and zero) one cannot help noticing certain simple but striking facts concerning these operations. The formal statements of these facts are called **natural laws.** These laws are of key importance in mathematics because all the procedures of arithmetic and algebra, and the associated reasoning, can be understood on the basis of a small number of such laws.

We have defined the **natural numbers** one, two, three, etc., as those number concepts associated with certain non-empty sets and we have defined **zero** as the number concept associated with the empty set. Since the laws we are going to discuss apply to zero as well as to the natural numbers, it will be convenient to have a name for the system containing *all* these numbers. We shall, therefore, adopt the name **counting numbers** to represent the system consisting of zero and all the natural numbers. (Some people prefer to call these "the **whole numbers**", altho it is somewhat difficult to recognize wholeness as a characteristic of zero.)

1.3.1. The Closure Laws

We define the meaning of the "addition" of two counting numbers to obtain a third one (called their sum) in the same way that we defined the addition of two natural numbers. From this it follows that the sum of any two counting numbers is, of course, also a counting number. We describe this state of affairs by saying that the system of counting numbers is "**closed** with respect to the operation of addition", or that the counting numbers are "closed under addition"; and this fact is referred to as the "**closure law** of addition for the counting numbers".

Since we have defined a product as a special kind of extended sum, it follows that the product of any two counting numbers is also a counting number; that is, **the system of counting numbers is closed under multiplication.**

Also, since a power is a special kind of extended product, it follows that the power formed with any counting number as base and any natural number as exponent is also a counting number. Thus **the counting numbers are closed under exponentiation** (if we exclude consideration of zero exponents, for which we have not yet defined a meaning).

Thus the system of counting numbers is closed with respect to each of the three natural operations—that is, the operations of addition, multiplication and exponentiation (zero exponents not being considered at present).

It may also be noticed that the system of natural numbers by itself is closed with respect to each of these three operations.

1.3.2. The Commutative Laws

1.3.2.1. *The commutative law for addition*

We recall that the addition of two to three is interpreted as follows. To a set with three elements there is adjoined a disjoint set with two elements and the required sum $(3 + 2)$ is defined to be the number of elements in the union so formed. Similarly, the addition of three to two may be interpreted as the determination of the number $(2 + 3)$ of elements in the union formed by adjoining a set of three elements to a disjoint set of two elements. The two unions described may be pictured as shown following on the left and right respectively.

Since the second set on the left and the first set on the right have the same number of elements (namely, two) it follows from the definition of number that the elements of these two sets can be put into one-to-one correspondence—for instance, by drawing

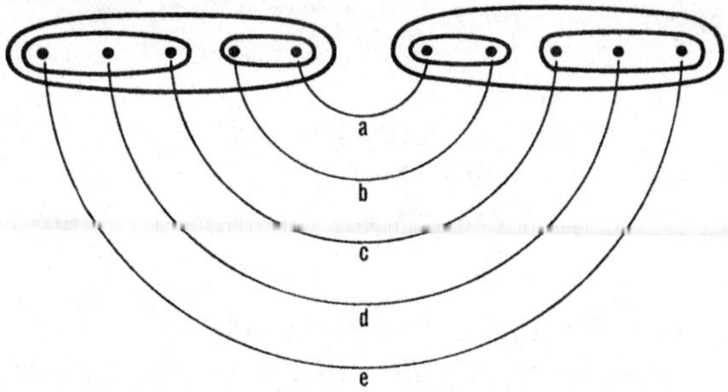

the lines labeled *a* and *b*. Similarly, since the first set on the left and the second set on the right have the same number of elements (namely, three) it follows that their elements can be put into one-to-one correspondence—as by drawing the lines labeled *c*, *d* and *e*. Now if we consider the aggregate of all the lines *a*, *b*, *c*, *d* and *e*, we notice that they show a one-to-one correspondence between the union on the left and the union on the right. This proves that the two unions have the same number of elements—that is, that the number 3 + 2 is the same as the number 2 + 3, or

$$3 + 2 = 2 + 3.$$

Notice that we have proved that these two numbers must be the same without actually finding the value of either one. Of course, by extended counting, we could show that 3 + 2 = 5 and also that 2 + 3 = 5, but the occurrence of the number five in both cases might, for all we know, be just a coincidence and we could not be sure on such a basis that 4 + 2 and 2 + 4, for instance, would be the same number. By disregarding the actual values of the sums in question and concentrating on the argument we have used, it can be seen that this same kind of argument would apply equally well to the sum of *any* two numbers taken first in one order and then in the other. On the basis of such an argument applied to four and two it is clear that 4 + 2 and 2 + 4 must necessarily represent the same number—whatever this number might be. Similarly, it could also be shown that

$$3 + 4 = 4 + 3$$
$$3 + 5 = 5 + 3$$
$$1 + 2 = 2 + 1$$
$$6 + 10 = 10 + 6$$
$$0 + 18 = 18 + 0$$
$$1{,}693{,}432 + 238{,}984 = 238{,}984 + 1{,}693{,}432,$$

and so on.

The general principle involved, that **the sum of two numbers is the same in whichever of the two possible orders they may be added,** or that the value of a sum of two numbers is not changed by reversing (or *commuting*) the order in which the two numbers are considered, is called the **commutative law for addition.** This is one of the important basic principles of arithmetic and algebra and it will be brought out later that this law for addition applies not just to the natural numbers but also to all number systems which we shall use.

1.3.2.2. *The use of literal numbers*

Having just discussed the commutative law for addition in words and having illustrated it by a particular example, this is a good time to introduce a new and more efficient way of stating this and other important mathematical principles—a method that is used very extensively to simplify and reduce the work of mathematical discussion.

Suppose we become weary of trying to write the commutative relation for sums of all possible pairs of natural numbers—which would be an endless task anyway—and suppose also that we wish to state the principle of commutativity of addition without becoming involved in a long, cumbersome, wordy explanation. We can state the principle very briefly but still with perfect completeness if we agree to just one new idea. Instead of talking about many specific natural numbers such as 1, 2, 3, 4, 5, 6, 10, 18, 238, 984, 1,693,432, etc., let us use non-numerical symbols such as, say, letters of the alphabet, to represent *some* natural numbers—*any* natural numbers—but ones which we shall not trouble to specify. For instance, we might think of the letters *a* and *b* as standing for some two natural numbers. (These *might* be 2 and 3, 2 and 4, 3 and 5, 1 and 2, 6 and 10, or *any* other numbers, but we shall not say which.) The commutative law for addition can now be expressed completely, precisely and very compactly by the following statement. **For any two natural numbers a and b, $a + b = b + a$.** If the manner in which the symbols a and b are used here to represent numbers is correctly understood, then this brief statement gives complete information on the commutative principle of addition in the natural numbers.

This method of stating mathematical principles or relations is called the method of **literal** symbols from the fact that letters are used to represent numbers, and the letters used in this manner are sometimes called **literal numbers.**

1.3.2.3. *The commutative law for multiplication*

The commutative law for multiplication is exactly similar to the commutative law for addition. Using the literal notation just discussed, we may state this new law in the simple form: **For any two natural numbers a and b, $a \cdot b = b \cdot a$.**

For instance, $3 \cdot 2 = 2 \cdot 3$, that is, three twos make the same number as two threes. To see why this should be true, picture the product $3 \cdot 2$ (remember this means the sum $2 + 2 + 2$) as represented by the union of three disjoint sets of two elements each, like this.

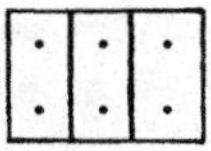

From this union we may use the first element of each of the three sets to form a set of three elements, and the second element of each of the three sets to form a second (and disjoint) set of three elements, like this.

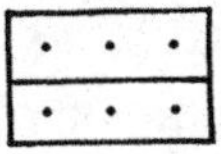

After forming these two new sets of three elements each, all the elements of the original union (of three sets of two elements each) have been used, and we have not used any other element. Since the same (union) set of elements is thus shown to constitute either the union of three sets of two elements each or the union of two sets of three elements each, it follows that the two unions can be put into one-to-one correspondence and, therefore, that they have the same number of elements. This justifies the statement that the number $3 \cdot 2$ is the same as the number $2 \cdot 3$.

In exactly the same way, for any two natural numbers a and b, we could associate the product $a \cdot b$ with the union of a sets of b elements each and the product $b \cdot a$ with the union of b sets of a elements each. We could then argue (by pointing out how either of these unions could be rearranged to form the other) that the two unions contain precisely the same elements and are therefore equal, and consequently that the numbers $a \cdot b$ and $b \cdot a$ are also equal.

1.3.3. The Associative Laws

1.3.3.1. *The associative law for addition*

In extending the idea of addition to more than two terms (or summands) it was pointed out that such addition involves a sequence of steps and that in each individual step *only two* numbers are actually added. That is, addition is essentially a *binary* (or *two*-number) operation. Thus in adding two, three and four, it was suggested that we first add three to two to get five and then add four to this five to get nine. Symbolically, this would be written as

$$\begin{aligned} 2 + 3 + 4 &= (2 + 3) + 4 \\ &= 5 + 4 \\ &= 9. \end{aligned}$$

If, on the other hand, we first add four to three to get seven and then add this seven to two, we also get the same number, nine, this way. The symbolic representation of this last procedure is

$$\begin{aligned} 2 + 3 + 4 &= 2 + (3 + 4) \\ &= 2 + 7 \\ &= 9. \end{aligned}$$

While the fact that we get the same final sum (nine) by either method of procedure is usually accepted by most people as a proper, suitable and happy sort of circumstance, the question might be asked whether this outcome should be attributed to coincidence or to inevitability. In other words, in adding any three natural numbers—say a, b and c—can we always be sure that the final sum obtained by first adding b to a to get $a + b$ and then adding c to this result to get what we write as

$$(a + b) + c$$

will be the same number as

$$a + (b + c)$$

which is the way we write the final result obtained by first adding c to b to get $b + c$ and then adding this to a? The answer is that the two final sums will always be the same, and its formal statement, **"For any three natural numbers a, b and c, $(a + b) + c = a + (b + c)$,"** is called the **associative law for addition** for natural numbers.

Why this is true can be seen by considering the sets and set operations associated with the numbers in question. For instance, in connection with the numbers $(2 + 3) + 4$ and $2 + (3 + 4)$, the first represents the total number of elements in a set formed thus: we first form the union of a set of two elements and a disjoint set of three elements, then we form the union of this union with a disjoint set of four elements; the second represents the total number of elements in a set formed thus: we form the union of a disjoint set of two elements with the union of a set of three elements and a disjoint set of four elements. Since we have a set of two elements entering into each of the two overall unions, the set of two elements in the first union can be put into a one-to-one correspondence with the set of two elements in

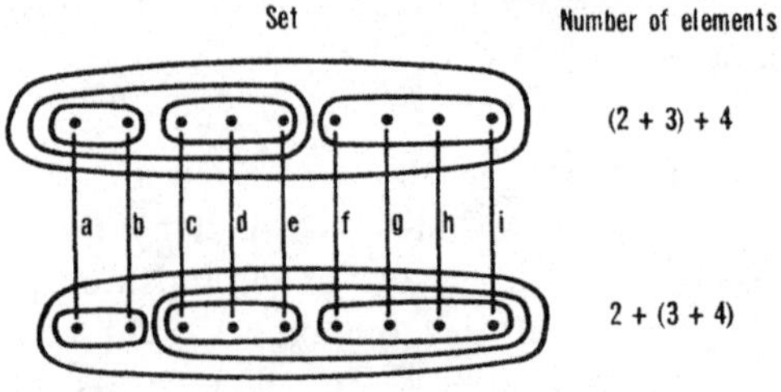

the second union. Similarly, the set of three elements in the first union can be put into one-to-one correspondence with the set of three elements in the second union, and the set of four elements in the first union can be put into one-to-one correspondence with the set of four elements in the second union. These one-to-one correspondences may be shown, in the usual manner, by drawing lines between corresponding elements—say lines (labeled) a and b in the case of the two-element sets, lines c, d and e in the case of the three-element sets, and lines f, g, h and i in the case of the four-element sets. Now considering the entire set of lines $\{a, b, c, d, e, f, g, h, i\}$, we see that they indicate a one-to-one correspondence between the elements of the first union and the elements of the second union. This proves that the number $(2 + 3) + 4$ of elements in the first union is the same as the number $2 + (3 + 4)$ of elements in the second union.

An argument of the same sort could, of course, be applied to the case of the numbers $(a + b) + c$ and $a + (b + c)$ where a, b and c are any natural numbers.

1.3.3.2. *The associative law for multiplication*

In extending the idea of multiplication to a product of three factors—such as $2 \cdot 3 \cdot 4$—we might interpret the product in either of two ways. We might think of first forming the product $3 \cdot 4$ (to get 12, of course) and then multiplying this resulting product by 2 (to get 24). To represent this way of thinking of the final product we write it as $2 \cdot (3 \cdot 4)$ and describe it by saying that, in it, the number three is *first associated* with the number four. Alternatively, we might think of multiplying the number four by the number (six) obtained by first forming the product $2 \cdot 3$. This latter way of regarding the final product is represented by writing $(2 \cdot 3) \cdot 4$, in which the number three is *associated first* with the number two. The fact that the final result obtained by either method is the same number (24) is not an accident or a coincidence. It is a fact that **for any three natural numbers a, b and c, $a \cdot (b \cdot c) = (a \cdot b) \cdot c$,** and the formal statement of this fact is what we call the **associative law for multiplication.**

As usual, a simple way to see why such a law is a true statement of fact is to consider the significance, with respect to unions of sets, of the numbers involved. For convenience in picturing the sets involved, let us consider sets of cube-shaped blocks like this.

If we think of a set of four blocks as arranged in a row like this,

then a set of $3 \cdot 4$ blocks (that is, the union of three disjoint sets of four blocks each) might be thought of as arranged in three rows of four blocks each, like this

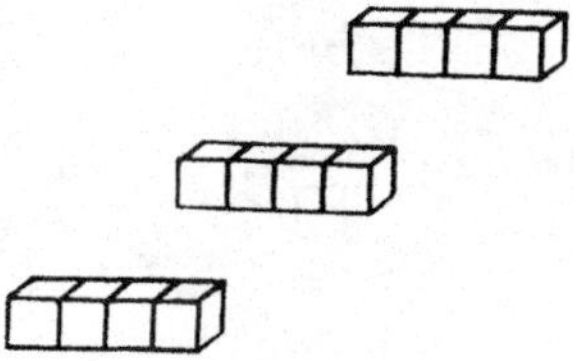

and, of course, we could also think of the three rows of blocks put together compactly like this.

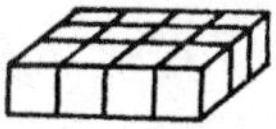

A set of $2 \cdot (3 \cdot 4)$ blocks could then be thought of as the union of two such layers of blocks arranged either like this

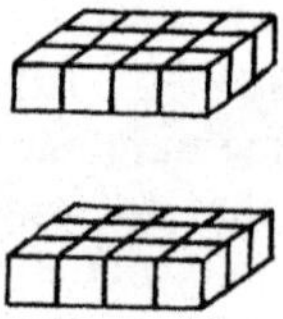

or, more compactly, like this.

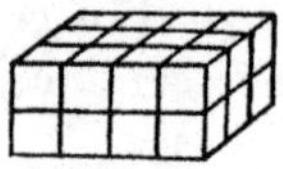

Now taking another viewpoint, we might first think of a set of three blocks arranged in a row like this;

a set of $2 \cdot 3$ blocks as arranged like this

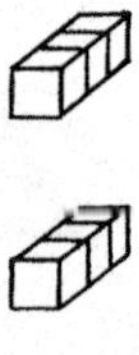

or, more compactly, like this;

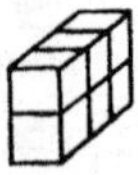

and a set of $(2 \cdot 3) \cdot 4$ blocks as

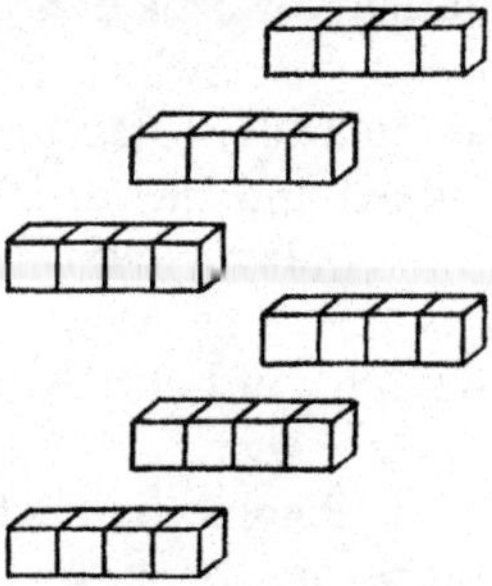

or, most compactly, like this.

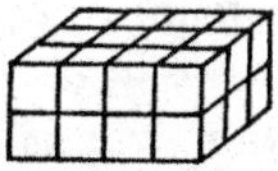

The point we are trying to make is that the numbers $2 \cdot (3 \cdot 4)$ and $(2 \cdot 3) \cdot 4$ both may be considered as the number of blocks in this latter set. Hence these two numbers must be the same, and it makes no difference whether we interpret the extended product $2 \cdot 3 \cdot 4$ as $2 \cdot (3 \cdot 4)$ or as $(2 \cdot 3) \cdot 4$, since

$$2 \cdot (3 \cdot 4) = (2 \cdot 3) \cdot 4.$$

Obviously, this same sort of argument could be used on any extended product $a \cdot b \cdot c$ (where the symbols a, b and c represent *any* three natural numbers) to show that

$$a \cdot (b \cdot c) = (a \cdot b) \cdot c.$$

This is our justification for the general law of associativity for multiplication.

1.3.4. The Distributive Law

We have discussed the natural laws of closure, commutativity and associativity for the operation of addition and the three similarly named laws for the operation of multiplication. Along with these, we have another law, called the **distributive** law, which explains the nature of the result when these two operations are used together. In particular it states that if any two natural numbers b and c are added and the resulting sum $b + c$ is then multiplied by any natural number a to form the product $a \cdot (b + c)$ the resulting number is the same as the one obtained when b and c are first individually multiplied by a to get the products $a \cdot b$ and $a \cdot c$ and these are then added to get the sum $a \cdot b + a \cdot c$. That is, it makes no difference in the final result whether the addition is performed before or after multiplication by the multiplier a. This principle is described by saying that multiplication by the number a can be *distributed* over each term of the sum $b + c$. Its formal statement is: **for any three natural numbers a, b and c, $a \cdot (b + c) = a \cdot b + a \cdot c$,** and it is called the **distributive law for multiplication over addition.**

For example, since

$$2 \cdot (3 + 4) = 2 \cdot 7 = 14$$

and

$$2 \cdot 3 + 2 \cdot 4 = 6 + 8 = 14,$$

it is seen that

$$2 \cdot (3 \cdot 4) = 2 \cdot 3 + 2 \cdot 4.$$

As with the other laws we have discussed, we can demonstrate the reasonableness of this law by considering sets associated with the numbers involved. If we represent sets of three and four elements respectively by

and

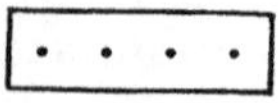

then a set of $3 + 4$ elements may be represented by

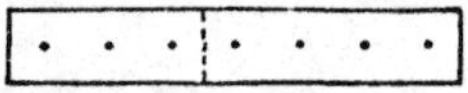

and a set of $2 \cdot (3 + 4)$ elements by the following figure.

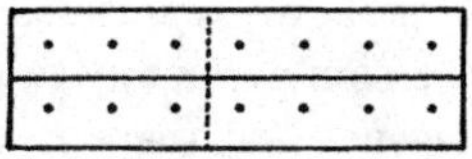

This, of course, is the same set as

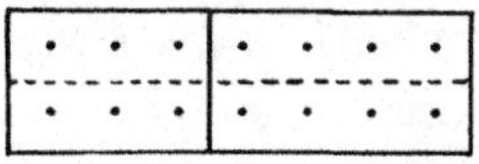

except for the way in which we consider the set as having been put together. The viewpoint suggested in the last diagram above is that of the union of a set of $2 \cdot 3$ elements with a disjoint set of $2 \cdot 4$ elements; hence the total number of elements in the union is also seen to be $2 \cdot 3 + 2 \cdot 4$. It then follows that

$$2 \cdot (3 + 4) = 2 \cdot 3 + 2 \cdot 4.$$

The same sort of argument could, of course, be applied to any three natural numbers a, b and c.

1.3.5. Identity Elements

We have already noted (on pages 36 and 37) that for any natural number n,

$$0 + n = n, \quad n + 0 = n,$$

and (on pages 39 and 40) that

$$1 \cdot n = n, \quad n \cdot 1 = n.$$

Because, as seen above, the addition of zero to any natural number yields a sum which is identical with the original number,

the number zero is called the identity element for the operation of addition.

Similarly, because the multiplication of any natural number by the number one yields a product which is identical with the original number, **the number one is called the identity element for the operation of multiplication.**

In general, if the symbol Δ represents some binary operation (that is, an operation for combining any two elements of some set S to produce a third element) and there is an element e in S such that for each element s in S it is true that

$$e \Delta s = s \text{ and } s \Delta e = s,$$

then e is called the identity element for the operation Δ in S. This concept is of considerable importance in advanced studies of algebra.

1.3.6. Use of Sentences in Describing Natural Laws and Other Relations

In discussing such relations as the associative, commutative and distributive laws we have described them in the words and sentences of ordinary prose, paying attention, to the best of our knowledge and ability, to the usual rules for clear and grammatical sentence structure. We have also illustrated these laws by means of examples and, finally, we have stated them in the precise and compact symbols common to mathematical usage.

It should be understood that in changing from expression in ordinary language to that of mathematical symbolism our object is to take advantage of greater precision and efficiency of expression—*not* to escape the rules of grammar. As a matter of fact, in order to achieve the precision and efficiency of expression for which mathematical symbols were invented they *must* be used with all the care and respect for propriety that goes into any linguistic composition—for if they are used carelessly the resulting symbolic statements will not be precise meaningful statements but meaningless gibberish.

Meaningful statements in mathematics—whether in standard prose or in symbols—must take the form of **sentences** with the same parts (subject, predicate, etc.) as in any other form of communication. In many mathematical statements the sentence structure is of the following very simple form. The verb form in the predicate is the word "equals" (or its representative symbol " = ") which may be translated to mean "is". Both subject and predicate nominative are numbers (either specific numbers or symbols which represent specific numbers).

As an example, the distributive law statement

$$a \cdot (b + c) = a \cdot b + a \cdot c$$

is a sentence with the following parts:

subject: "the number (represented by) $a \cdot (b + c)$"
verb: "is"
predicate nominative: "the number (represented by) $a \cdot b + a \cdot c$".

(Due to the meaning of the verb "equals", of course, the subject and predicate nominative may be interchanged in any sentence in which "equals" is the verb.)

It is important to notice that **while**

$$\boldsymbol{a \cdot (b + c) = a \cdot b + a \cdot c}$$

is a complete and proper sentence, and tells exactly how $a \cdot (b + c)$ and $a \cdot b + a \cdot c$ are related, any mere listing of $a \cdot (b + c)$ and $a \cdot b + a \cdot c$ such as

$$a \cdot (b + c) \quad a \cdot b + a \cdot c,$$

or

$$a \cdot (b + c), a \cdot b + a \cdot c,$$

or

$$a \cdot (b + c)$$
$$a \cdot b + a \cdot c, \text{ etc}$$

does not constitute a proper sentence at all since the verb form is missing, **and conveys no usable information.**

It is also important to remember that *no other symbol*—such as a dash, an arrow, a double-ended arrow, a tilde, etc.—may legitimately be substituted for the equality symbol since no other symbol has the same meaning. Thus **while**

$$a \cdot (b + c) = a \cdot b + a \cdot c$$

is a meaningful sentence, none of the following is either meaningful or a sentence.

$$a \cdot (b + c) - a \cdot b + a \cdot c$$
$$a \cdot (b + c) \rightarrow a \cdot b + a \cdot c$$
$$a \cdot (b + c) \longleftrightarrow a \cdot b + a \cdot c$$
$$a \cdot (b + c) \sim a \cdot b + a \cdot c$$

Since it is obvious that writing meaningless marks on paper accomplishes nothing and is a complete waste of time and effort, care should be taken to use both words and symbols only in a way that yields a clear and useful meaning.

In order to avoid arriving at false conclusions in one's reasoning it is also necessary to refrain carefully from making any statement which is not known to be true. To illustrate the point, suppose we wish to verify the truth of the statement

$$a \cdot (b + c) = a \cdot b + a \cdot c$$

for the case in which $a = 2$, $b = 3$ and $c = 4$. To start by writing

$$2 \cdot (3 + 4) = 2 \cdot 3 + 2 \cdot 4$$

would be absurd since it is the truth of this statement which is being questioned. If we know the statement to be true, then our task is completed and there is no point in discussing the matter further. If we do not know whether the statement is true or false then certainly our ignorance on this point cannot be taken as justification for writing the statement as tho it were true.

The straightforward way to determine whether or not the statement

$$a \cdot (b + c) = a \cdot b + a \cdot c$$

is true if $a = 2$, $b = 3$ and $c = 4$ is to find the value which each of the expressions $a \cdot (b + c)$ and $a \cdot b + a \cdot c$ has if $a = 2$, $b = 3$ and $c = 4$. A comparison of these values will then make clear immediately whether or not the statement is true in this case. This procedure is illustrated by the following.

If $a = 2$, $b = 3$ and $c = 4$, then

$$\begin{aligned} a \cdot (b + c) &= 2 \cdot (3 + 4) \\ &= 2 \cdot 7 \\ &= 14 \end{aligned}$$

and

$$\begin{aligned} a \cdot b + a \cdot c &= 2 \cdot 3 + 2 \cdot 4 \\ &= 6 + 8 \\ &= 14. \end{aligned}$$

Since the value of $a \cdot (b + c)$ and the value of $a \cdot b + a \cdot c$ are both seen to be the same number, 14, in this case, it then follows clearly that if $a = 2$, $b = 3$ and $c = 4$, then it can be said truthfully that

$$a \cdot (b + c) = a \cdot b + a \cdot c.$$

Notice carefully that in this way of discussing the problem we have not at any time written any statement other than those which we *know* to be *true*.

Exercises

1. Is exponentiation commutative in the system of natural numbers? Justify your answer. (Hint: In investigating the matter of whether the natural laws apply to exponentiation, it is helpful to replace the customary non-symmetric exponential notation, a^b, by a notation with left-right symmetry of appearance such as, say, $a * b$.)

2. Is exponentiation associative in the system of natural numbers? Justify your answer.

3. Is the commutative law of addition valid in the system of counting numbers? Justify your answer by providing the necessary supplement to the discussion already given for the natural numbers.

4. Modify Exercise 3 to apply to the associative law of addition, and answer the question as modified.

5. Modify Exercise 3 to apply to the commutative law of multiplication, and answer the question as modified.

6. Modify Exercise 3 to apply to the associative law of multiplication, and answer the question as modified.

7. We say that "multiplication is distributive over addition" in the system of natural numbers because for any three natural numbers a, b and c it is true that $a \cdot (b + c) = (a \cdot b) + (a \cdot c)$. Since it *cannot* be said truthfully that "for any three natural numbers a, b and c it is true that $a + (b = c) = (a + b) \cdot (a + c)$", we say that "addition is *not* distributive over multiplication". Is exponentiation distributive over addition in the natural number system? Justify your answer.

8. Is exponentiation distributive over multiplication in the natural number system? Justify your answer.

9. Is addition distributive over exponentiation in the natural number system? Justify your answer.

10. Is multiplication distributive over exponentiation in the natural number system? Justify your answer.

11. Is there an identity element for exponentiation? Justify your answer.

12. Give an example of a binary operation (other than those already discussed) defined on some set of elements (not necessarily numbers) for which there is an identity element.

13. Draw a diagram and give the accompanying explanation to illustrate the principle of additive commutativity as on page 47 but using different numbers.

14. Draw diagrams and give the accompanying explanation to illustrate the principle of multiplicative commutativity as on page 49 but using different numbers.

15. Draw a diagram and give the accompanying explanation to illustrate the principle of additive associativity as on page 52 but with different numbers.

16. Draw diagrams and give the accompanying explanation to illustrate the principle of multiplicative associativity as on pages 52 thru 55 but with different numbers.

17. Draw diagrams and give the accompanying explanation to illustrate the principle of distributivity of multiplication over addition as on pages 56 and 57 but with different numbers.

18. Give a careful demonstration of the use of mathematical *sentences* to verify that *if* $m = 7$, $n = 2$ and $p = 6$, *then*:

(a) $m + n = n + m$, (b) $m + (n + p) = (m + n) + p$,
(c) $m \cdot n = n \cdot m$, (d) $m \cdot (n \cdot p) = (m \cdot n) \cdot p$,
(e) $(m + n) \cdot p = m \cdot p + n \cdot p$.

19. Modify and work Exercise 18 for the case in which $m = 0$, $n = 0$ and $p = 0$.

CHAPTER 1.4

THE INVERSE OPERATIONS AND CONVERGENCE; THE NUMBER SYSTEMS GENERATED BY THESE OPERATIONS

1.4.1. The Nature of an Inverse Operation

"There was a man in our town
And he was wondrous wise.
He jumped into a bramble bush
And scratched out both his eyes.

When his eyes were open and he saw
He could not see—why then
He jumped back in the bramble bush
And scratched them in again."

This resourceful action on the part of a nursery-rhyme character illustrates the nature of an inverse operation. Having scratched his eyes out, he found a way of replacing them and thus of restoring himself to his original condition. Any action or operation which has this effect of counteracting or neutralizing the effect of a previous operation is called the **inverse** of that previous operation.

The armed forces recognize the usefulness of the inverse operation concept by authorizing the command "As you were!" to be given whenever a maneuver is improperly executed. If, for example, some of the recruits march to the left when they should march to the right, the command "As you were!" returns them all to their starting positions ready to start all over again. This device frequently saves the day on the parade ground altho it is of limited applicability in combat situations.

We shall find that there is an inverse operation for each of the natural operations of addition and multiplication and that the operation of exponentiation has two inverses. We shall also find that the use of these inverse operations leads to the development of our system of numbers from the counting numbers to more extensive and intricate systems.

1.4.2. Subtraction; the Integers

1.4.2.1. *Subtraction as the inverse of addition*

The general problem leading to the operation of addition is: "If two numbers (say, a and b) are given, what number is the sum $a + b$?" That is, the problem calls for the finding of a new number (say c) such that

$$a + b = c.$$

In view of the commutativity of addition this is, of course, the same as the problem of finding the number c such that

$$b + a = c.$$

For example, if a is 2 and b is 3, then the related addition problem is to find the number 5 since

$$2 + 3 = 5 \text{ and } 3 + 2 = 5.$$

The inverse to the problem of addition is to be given a sum and one of its summands and be required to find the other summand uniquely. That is, if we are given a number (say c) as a sum and a number (say a) as one of the summands, then we are to find a number, and *just one* such number (say b) such that

$$a + b = c \text{ (and, of course, also } b + a = c\text{)}.$$

For example, if we are told that a sum is 5 and one of the summands is 2 and we are asked to find the other summand, this means that we must find some number (say b) such that

$$2 + b = 5 \text{ and } b + 2 = 5.$$

If we remember the facts of addition, it is probably fairly obvious that b must be the number 3 in this case.

In working with the general problem of finding a summand b which, when added to a summand a, yields a sum c, the required summand b (whether its actual identity as a particular number is known or not) is customarily represented by the symbol

$$c - a.$$

Thus we are *defining* this symbol as representing a number for which it is true that

$$a + (c - a) = c$$

and

$$(c - a) + a = c.$$

The number represented is called a **difference** and the symbol is read as "c minus a". The process of determining or identifying a difference is called **subtraction** (or also, of course, **inverse addition**). It should be noted that subtraction is a binary operation.

For example, we write

$$5 - 3 = 2$$

because of the fact that

$$3 + 2 = 5 \text{ (and } 2 + 3 = 5),$$

and we call the process of identifying the difference $5 - 3$ (as the number two) the "subtraction of three from five".

1.4.2.2. *Subtraction and the natural laws*

Having encountered the new (inverse) operation of subtraction it is natural to wonder if the same natural laws which applied to the three natural binary operations also apply to this new binary operation. The answer to this question can be found by examining a few simple examples such as the following.

Since, for instance, there does not exist any natural number b such that

$$5 + b = 3$$

it is seen immediately that the difference $3 - 5$ (if it exists in any sense) is not a natural number. Hence, **the natural numbers are not closed under subtraction.**

Since $5 - 3$ is a natural number (namely, two) and $3 - 5$ is not any natural number at all, it is clear that these two differences cannot be equal. Hence this example shows that **subtraction is not commutative.**

We say that "addition of natural numbers is associative" to describe the fact that for any three natural numbers a, b and c the two extended sums $(a + b) + c$ and $a + (b + c)$ are equal. Analogously, we might say that "subtraction of the natural numbers is associative", or not, according as the two extended differences $(a - b) - c$ and $a - (b - c)$ are, or are not, generally equal. For the case with $a = 7$, $b = 4$ and $c = 2$, it is seen that

$$(a - b) - c = (7 - 4) - 2 = 3 - 2 = 1$$

and

$$a - (b - c) = 7 - (4 - 2) = 7 - 2 = 5.$$

This single example proves that **subtraction is not associative.**

We say that multiplication is distributive over addition because for any three natural numbers a, b and c it is true that $a \cdot (b + c) = a \cdot b + a \cdot c$. By analogy, we shall say that multiplication is distributive over subtraction if, for any three natural numbers a, b and c, it is true that

$$a \cdot (b - c) = a \cdot b - a \cdot c.$$

In testing the truth of this statement we recall that, due to the way in which we defined a difference, this statement will be true if, and only if,

$$a \cdot (b - c) + a \cdot c = a \cdot b,$$

or, making use of the distributivity of multiplication over addition, if

$$a \cdot [(b - c) + c] = a \cdot b.$$

But, by the definition of a difference, it is true that

$$(b - c) + c = b$$

and hence that

$$a \cdot [(b - c) + c] = a \cdot b$$

and, therefore, that

$$a \cdot (b - c) = a \cdot b - a \cdot c.$$

Thus **multiplication is distributive over subtraction.** (We have given the proof here for the case of "multiplication on the left". The proof would be similar for the case in which the common multiplier a is on the right.)

If there were to be an identity element (say e) for subtraction in some system of numbers, then it would have to be true that for each number (say a) in that system,

$$a - e = a \text{ and } e - a = a.$$

It is true that in the system of counting numbers we have the number zero and that for each counting number a

$$a - 0 = a$$

because of the fact that

$$a = 0 + a.$$

However, there is no counting number e such that (for every counting number a)

$$e - a = a.$$

Hence **there is no identity element for subtraction in the system of counting numbers.**

1.4.2.3. *The relations "greater than" and "less than"*

On page 31 we indicated in a rudimentary way the meaning

associated with the word "greater" and the expression "greater than". Here we present this same concept (and the accompanying concept of "less than") in terms of the operation of addition.

We have noticed that, starting with the two natural numbers 3 and 5, it is possible to find a natural number b (namely 2, of course) such that

$$3 + b = 5$$

but it is impossible to find any natural number d such that

$$5 + d = 3.$$

In general, we may discover that for any two *different* natural numbers a and c, if it is possible to find a natural number b such that

$$a + b = c,$$

then it is *not* possible to find any natural number d such that

$$c + d = a$$

and, conversely, if it is not possible to find a natural number d such that

$$c + d = a,$$

then it *is* possible to find a natural number b such that

$$a + b = c.$$

Whenever this situation exists, we describe it by saying that "c is **greater than** a" and "a is **less than** c" or, symbolically,

$$c > a \text{ and } a < c$$

respectively.

Both "greater than" and "less than" are examples of what are called **order relations.**

It will be noticed that if, for two counting numbers a and c, $c > a$, then a can be subtracted from c and the resulting difference $c - a$ will be found among the natural numbers, and that such will not be the case if $c < a$.

1.4.2.4. *Enlargement of the number system to provide closure for subtraction; the integers*

An "apocryphal story" is, roughly speaking, a story, not necessarily true but usually plausible, about how something *might* have happened. As an example, we offer the following.

Once upon a time in a country called Posit the children were accustomed to studying their arithmetic by playing a game in which counting numbers were represented by pacing off certain distances toward the east. Starting at the international boundary line a distance of two paces to the east represented the number two; two paces to the east followed by three additional paces to the east represented the sum $2 + 3$; three paces to the east followed by two more represented $3 + 2$; etc. By playing this game the children were enabled to make visual comparisons of the sizes of various numbers, to satisfy themselves that the commutative law of addition actually worked for such simple cases as $2 + 3 = 3 + 2$ and that the associative law of addition worked for similarly simple cases, and to illustrate the idea of subtraction. In particular, the Positive children (as they were called) had convinced themselves that it was impossible to subtract five from three; that is, to find a number which could be added to five to yield a sum of three.

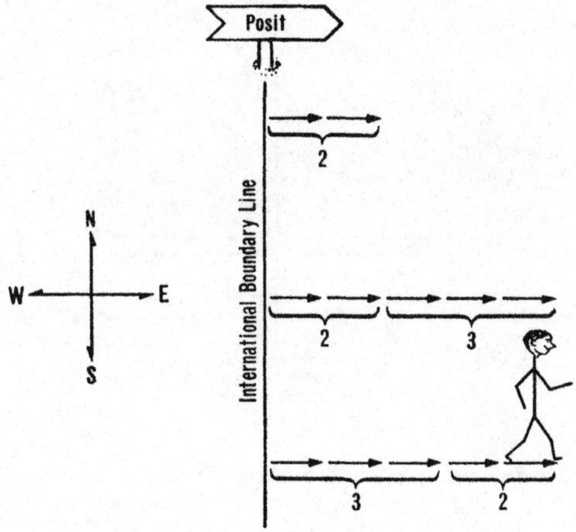

Occasionally, a foreign boy called Negat would attempt to join the Positive children in their game or, being repulsed, would play his own version of their game on his own side of the boundary line. Being a foreigner, he did many things backward, and it was a source of great amusement to the Positive children that he represented numbers by pacing off distances from the boundary line toward the *west* rather than toward the east in the approved fashion. Since the Positive children recognized only eastward distances as representing numbers they refused to recognize Negat's pacings as representing proper numbers and referred to them jokingly as "negative numbers".† To emphasize their pride in their own national way of doing things, they also began to refer to their eastward-measured numbers as "positive numbers". Negative numbers were written as "n1", "n2", "n3", etc. In spite of the unconventionality of Negat's number representations, it was noticed that as long as he stayed on his own side of the boundary line all the known principles of arithmetic seemed to work out as well with the negative numbers as they did with the numbers of the Positive children. Due partly to this fact and partly to a growing spirit of tolerance among the Positive children,

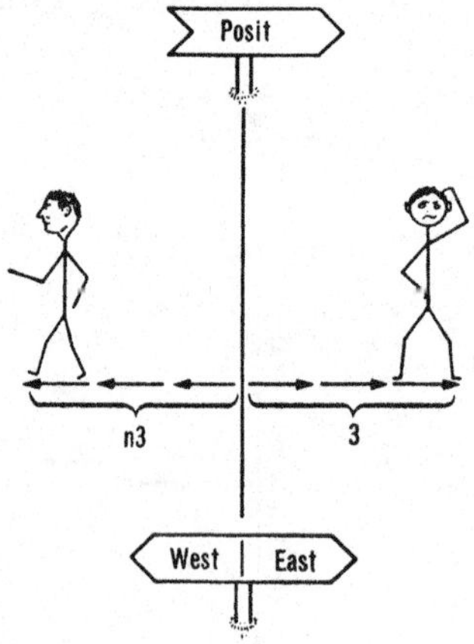

† Even to this day, when someone does something in a manner contrary to local custom, it is sometimes said that the person has "a negative attitude".

Negat and his numbers were finally accepted in the Positive children's games along with the assumption that all the known natural laws held for the negative numbers as well as the positive numbers.

With Negat playing and using his numbers on the Positive side of the boundary it soon became apparent that five could be subtracted from three and that, as a matter of fact,

$$3 - 5 = \text{n}2$$

since it could be seen to be true that

$$5 + \text{n}2 = 3 \text{ and } \text{n}2 + 5 = 3.$$

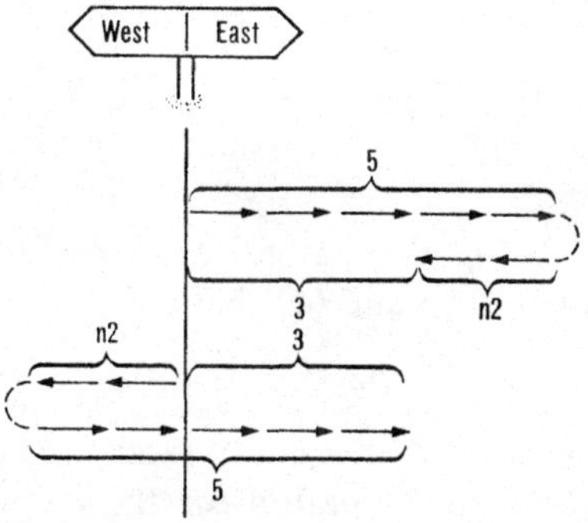

It is interesting to note that this result can be verified not only thru the pacing game but also as a consequence of the assumption of the validity of the natural laws for the negative numbers, thus:

$$\begin{aligned} 5 + \text{n}2 &= (3 + 2) + \text{n}2 \\ &= 3 + (2 + \text{n}2), \text{ by the associative law for addition,} \\ &= 3 + 0 = 3. \end{aligned}$$

Since zero paces either east or west from the international boundary line leaves one on that line, it is clear that a position on that line represents the number zero. Since any number of paces either east or west from this starting line followed by the same number of paces in the opposite direction also leaves one on this line it follows also that all numbers such as 2 + n2, n2 + 2, 3 + n3, n3 + 3, 5 + n5, n5 + 5, etc., are also represented by a position on this line. Thus, according to our pacing game model of numbers and addition, it is indicated that

$$2 + n2 = 0, n2 + 2 = 0, \text{etc.}$$

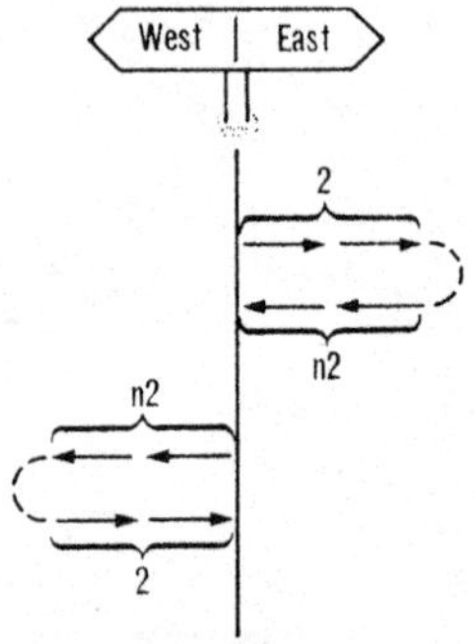

It then follows, using the natural analog of our definition of "difference" on pages 64 and 65, that

$$n2 = 0 - 2.$$

As a consequence of this equality (and similar equalities such as n3 = 0 − 3, etc.) the notation for the number n2 soon was changed to a combination of the notations n2 and 0 − 2 and was thenceforth written as −2. Similarly, n3 became −3, n4 became −4, etc. (Some people prefer to write $^-2$, $^-3$, $^-4$, etc.)

In terms of this new notation the equality

$$5 + n2 = 3$$

would now be written as

$$5 + (-2) = 3.$$

Comparing this with the equality

$$5 - 2 = 3,$$

it is seen that the addition of −2 and the subtraction of 2 have exactly the same effect. In the same way it can be seen that **the addition of any negative number, say $-b$, is the same as the subtraction of the corresponding positive number b.** It should be noted here that our reasoning is essentially based on the fundamental relations:

$$b + (-b) = 0, \quad (-b) + b = 0.$$

To examine into the meaning of the subtraction of a negative number we first recall the manner in which we defined subtraction of a positive number. For positive numbers a, b and c, we say

$$a = c - b$$

if, and only if,

$$a + b = c$$

and a is the *only* number which is related to b and c in this manner. If we define the concept of subtraction in eactly the same way for the subtraction of a negative number, then we shall say

$$x = z - (-y)$$

if, and only if, x is that unique number for which

$$x + (-y) = z.$$

But since we have agreed that $x + (-y)$ and $x - y$ have the same meaning, this means that

$$x = z - (-y)$$

if, and only if, x is that unique number for which

$$x - y = z,$$

and this latter statement means that

$$x = z + y.$$

Hence the statements

$$x = z - (-y)$$

and

$$x = z + y$$

are equivalent. In words, this means that **the subtraction of a negative number is the same as the addition of the corresponding positive number.**

If, in the preceding discussion, we consider the special case n which the symbol z stands for the number zero, the result shows that the difference $0 - (-y)$ is the same number as $0 + y$, or simply y. Also, recalling that we have identified the number $-b$ ("negative b") with the difference $0 - b$, it is now consistent to identify the number $-(-y)$, which we call "the negative of negative y", with the difference $0 - (-y)$. The result then is that

$$-(-y) = y$$

or, in words, that **the negative of a negative number is the corresponding positive number.**

At about this time it would not be abnormal or even unusual for someone to ask: "Just what (of any importance) does an imaginary game played by imaginary children have to do with the real and serious business of working with numbers? Why should we consider that such things as negative numbers really exist merely because the addition or subtraction of them can be likened to the combination of certain pacings west or east?"

The answer to this is that *if* we *define* negative numbers -1, -2, -3, etc., by saying that they are numbers of such a nature that

$$(-1) + 1 = 0 \text{ and } 1 + (-1) = 0,$$
$$(-2) + 2 = 0 \text{ and } 2 + (-2) = 0,$$
$$(-3) + 3 = 0 \text{ and } 3 + (-3) = 0, \text{ etc.},$$

then by pointing out the existence of a simple and easily understood interpretation of such concepts in the physical world (the westward pacings of Negat as contrasted with the eastward pacings of the Posit children) we are giving actual proof that such concepts as **positive and negative numbers can and do exist in relation to observable physical realities.** Thus, in spite of the fact that negative numbers exist (strictly speaking) only in our imagination, we should not consider them as in any way inferior to the positive numbers since, as a matter of fact, **it is only in our mental awareness of them that any numbers at all actually exist.**

While we cannot say with certainty that the story of Negat is *not* true in every detail and particular, it is highly probable that such a story is, indeed, not completely true in all respects since it is apocryphal and we concocted it only for the purpose of presenting a simplified picture of how, in general, men have groped their way toward the formulation of the concepts of negative numbers and negativity. It is also highly probable, however, that the actual overall experience of thinking mankind† in formulating these concepts was of a nature so generally similar to the apocryphal one described that, as far as the resulting mathematical concepts and their use are concerned, the whole development could just as well have occurred as we have fictitiously described it.

1.4.2.4.1. *Multiplication with the integers.* Just as the symbol 3 represents the number concept three only because we agreed to recognize it as doing so (we could just as well have chosen instead the symbol III as the ancient Romans did, or the symbol ٣ as the Arabs did and still do), in the same way any symbol has only that meaning which is agreed on as appropriate by those who use it. For example, the symbol $4 \cdot 3$ means the sum $3 + 3 + 3 + 3$ only because we have agreed to accept the symbol as representing the specified sum. Note that, up to the time this was done, the symbol $4 \cdot 3$ had no meaning at all.

Consequently, when we look for a meaning for such symbols as $4 \cdot (-3)$, $(-4) \cdot 3$ and $(-4) \cdot (-3)$ we must understand that since no meaning has so far been assigned to them by any previous definition, they have no meaning at all at present and we may use them to represent any meaning we wish as long as the meanings we assign to them do not result in anything in contradiction with previously adopted usages. Subject only to this limitation, we are free to assign whatever meaning to these symbols we consider to be most appropriate, useful or desirable.

† Thinking mankind—neat example of a proper subset.

Since we chose to define $4 \cdot 3$, for instance, as meaning $3 + 3 + 3 + 3$ or 12, it is most natural to define $4 \cdot (-3)$ as $(-3) + (-3) + (-3) + (-3)$ which, of course, is -12. Since this seems such a natural thing to do, and since we can think of no reason (even after protracted effort) for not doing so, we do it. Thus, *as a matter of definition*,

$$4 \cdot (-3) = (-3) + (-3) + (-3) + (-3) = -12$$

and hence

$$4 \cdot (-3) = -(4 \cdot 3).$$

Similarly, **for any two natural numbers *a* and *b*, we define**

$$\boldsymbol{a \cdot (-b) = -(a \cdot b).}$$

The preceding definition required little effort since it is about as easy to think of four negative threes as it is to think of four ordinary, or positive, threes. However, since it is somewhat more difficult to think of "negative four" threes, it is not as simple to find a suitable meaning for $(-4) \cdot 3$.

Remembering that multiplication of the natural numbers was simplified somewhat by the existence of the commutative law for multiplication, it occurs to us that it would be desirable, if possible, to have this law valid for *all* products. In accordance with this thought, we *define*

$$(-4) \cdot 3 = 3 \cdot (-4).$$

But, by our previous definition,

$$3 \cdot (-4) = (-4) + (-4) + (-4) = -12 = -(3 \cdot 4) = -(4 \cdot 3)$$

so that

$$(-4) \cdot 3 = -(4 \cdot 3).$$

Hence, as a consequence of our definition,

$$(-4) \cdot 3 = 4 \cdot (-3).$$

Similarly, **for** the general case of **any two natural numbers** a **and** b **we define**

$$(-a) \cdot b = b \cdot (-a)$$

as a consequence of which we have

$$(-a) \cdot b = -(a \cdot b)$$

and also

$$(-a) \cdot b = a \cdot (-b).$$

In defining a meaning for the still undefined concept of such products as $(-4) \cdot (-3)$, we may again be guided by considerations of what would be most desirable. Having already based our definition of products of the type of $(-4) \cdot 3$ on a desire to extend the validity of the commutative law of multiplication to our newly expanded number system, it seems most natural to design further definitions in such a way that all of the already familiar laws will remain valid in the new number system, if this is possible. Therefore, let us *assume* that the usual commutative, associative and distributive laws are valid in our new number system (consisting of the positive integers, zero and the negative integers) and see what meaning this assumption leads us to accept for products such as $(-4) \cdot (-3)$. We reason that

$(-4) \cdot (-3) = (-4) \cdot (-3) + 0$, due to the nature of zero,
$= (-4) \cdot (-3) + 0 \cdot 3$, due to the nature of zero,
$= (-4) \cdot (-3) + [(-4) + 4] \cdot 3$, by the definition of -4,
$= (-4) \cdot (-3) + [(-4) \cdot 3 + 4 \cdot 3]$, by the distributive law,
$= [(-4) \cdot (-3) + (-4) \cdot 3] + 4 \cdot 3$, by the associative law for addition,
$= (-4) \cdot [(-3) + 3] + 4 \cdot 3$, by the distributive law,
$= (-4) \cdot 0 + 4 \cdot 3$, by the definition of -3,
$= 0 + 4 \cdot 3$, due to the nature of zero,
$= 4 \cdot 3$, due to the nature of zero.

Thus, on the basis of our assumption of the continued validity of the fundamental laws, we have no choice but to define $(-4) \cdot (-3)$ to mean $4 \cdot 3$. Similarly, **for any natural numbers *a* and *b*, we define**

$$(-a) \cdot (-b) = a \cdot b.$$

As a matter of fact, it is not even necessary to restrict the symbols a and b to represent just natural numbers, since the kind of argument used above would be equally valid if each of the symbols involved were to represent *any natural number, or zero, or any negative natural number.*

To summarize what we have learned from watching the imaginary Posit children at play with their friend Negat (we no longer regard him as an odd foreigner but consider him now as a friend), we see that we can construct an enlarged system of numbers which contains the previously known natural numbers and zero and also a new set of "negative" numbers which are very similar to, but in a certain sense "opposite" to, the natural numbers. This enlarged system has been constructed in such a way as to make it **closed for the operation of subtraction.** By defining (implicitly) the nature of "addition" for the new numbers in terms of the pacing game we were able to see intuitively that the customary laws of addition were valid. By deciding to *make* the laws for multiplication valid in the new system, we thereby determined the necessary nature of "multiplication" in the new system. We are now ready, therefore, to perform the operations of addition, subtraction and multiplication with any numbers of the enlarged system.

To celebrate the completion of the construction of this enlarged number system and the integration of Negat and his negative numbers into the community of numbers, we shall call all the numbers in the enlarged system **integers,** calling the original natural numbers **positive integers** from now on, and using the designation **negative integers** to describe Negat's backward version of the natural numbers. Notice that **zero is also an integer altho neither positive nor negative.**

EXERCISES

1. There are many examples in everyday life of operations which have inverses. For example, locking and unlocking a door are operations each of which is the inverse of the other; so are driving a car into a garage and backing it out. (a) Describe five such operations and their inverses. (Be original!) (b) Describe five operations which do not have (in a practical sense) any inverse.

2. Perform the indicated additions and subtractions, illustrating each case by drawing a figure in which the addition or subtraction of each number is represented by a line segment of appropriate length drawn to the left or right.

EXAMPLE

$$\begin{aligned}(-3) + 5 - (-2) - 1 &= [(-3) + 5] - (-2) - 1\\ &= 2 - (-2) - 1\\ &= [2 - (-2)] - 1\\ &= [2 + 2] - 1\\ &= 4 - 1\\ &= 3.\end{aligned}$$

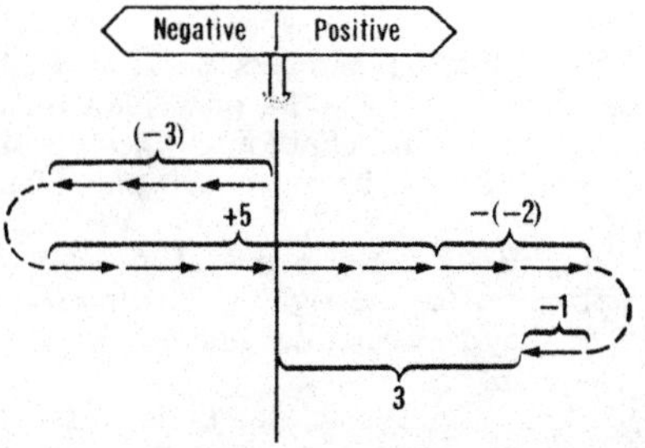

(a) $3 + 4 - 5$ (b) $0 - (-3)$ (c) $3 - 3 + (-3) - (-3)$
(d) $-7 - (-2) + (-5) + 3$ (e) $2 + (-6) - (-3) - 5 - (-2)$.

3. Evaluate each of the following expressions and illustrate the process with an accompanying figure, as in Exercise 2.

(a) $-3 + 6 + (-4) - (-2)$ (b) $2 - 7 - (-3) + 6$
(c) $1 - 2 + (-3) - (-4)$ (d) $2 + 2 - 2 + (-2) - (-2)$.

4. Explain what meanings are to be associated with the symbols: (a) -0; (b) $-(-5)$. (Try to *base your explanation as directly as possible on the definitions and properties which we have formally recognised* in connection with such concepts as "zero", "negative numbers", "subtraction", etc.)

5. Carry out the indicated subtractions and multiplications.

(a) $-2 -5 -4$ (b) $-2 - 5 \cdot (-4)$ (c) $(-2 - 5) \cdot (-4)$
(d) $-(2 - 5) \cdot (-4)$ (e) $(-2) \cdot (-5) \cdot (-4)$.

6. Carry out the indicated operations.

(a) $-3 - 6 - 8$ (b) $-(3 - 6) - 8$ (c) $-3 - (6 - 8)$
(d) $(-3 - 6) \cdot (-8)$ (e) $-(3 - 6 - 8)$.

7. Formulate and state a rule for determining whether an extended product, involving a specified number (say p) of positive factors and a specified number (say q) of negative factors, is a positive number or a negative number.

8. Is there an inverse for the operation of multiplication? Explain.

9. Is there an inverse for the operation of exponentiation? Explain.

10. Is there an inverse for the operation of subtraction? Explain.

11. Give three examples of quantities whose measurement may be regarded as involving an application of negative numbers.

12. Construct a multiplication table showing the product of each number in the set $\{-10, -9, -8, -7, -6, -5, -4, -3, -2, -1, 0, 1, 2, 3, 4, 5, 6, 7, 8, 9, 10\}$ by each number in this set.

13. To serve his purpose in a particular discussion, Dr. Hoho *defines* a "flageomizer" as "any pair of urmfs which is not a snafflemeyer". Criticize this definition from the point of view of mathematical propriety. (In discussing this problem, the terms "urmf" and "snafflemeyer" are to be treated as terms whose meaning is well understood by everyone concerned.)

14. Using the definition stated in section 1.4.2.3 for "greater than" and "less than" among the natural numbers as a model, compose a suitable similar definition for these same two relations among the integers.

15. State both symbolically and in words the order relation between each pair of the following set of natural numbers: $\{6, 2, 14, 17, 1, 97, 5\}$. (*Example*: "$6 > 2$, or 6 is greater than 2", "$6 < 14$, or 6 is less than 14", etc.)

16. State (on the basis of your definition in Exercise 14) both symbolically and in words the order relation between each pair of the following set of integers: $\{6, -2, 14, -17, -1, -97, 0, 5\}$.

17. Without making use of the principle of multiplicative commutativity, prove that "multiplication on the right is distributive over subtraction"; that is, prove that for any three natural numbers a, b and c, it is true that $(b - c) \cdot a = b \cdot a - c \cdot a$.

18. Is the set of negative integers closed under (a) addition? (b) multiplication? (c) subtraction?

1.4.3. Division; the Rational Numbers

1.4.3.1. *Division as the inverse of multiplication*

We have presented the notion of the negative integers as "opposites" of the natural numbers (or positive integers) and we have represented the origin of the concept of inverse addition (or subtraction) as being prompted by a desire to make the solution of certain inverse problems in addition possible. This, of course, is only a very roughly true historical picture and it has been presented only to make possible an easy intuitive appreciation of the concepts.

Actually, the development of the concepts of subtraction and the negative numbers probably occurred in response to needs encountered in connection with practical human problems and desires. Ever since human beings first began collecting clams, sheep, etc., there has been a need to answer such questions as: "How far will I be from home if I first travel nine hollers to the north, then four hollers to the south, and then thirteen hollers to the north?"; "How many clams should I pay the landlord if I owe him thirty clams for the rent of the cave and he owes me seven clams for the fish I gave him?"; etc. To answer such questions it is convenient and helpful, as a practical matter, to have such concepts as subtraction and negative numbers available.

For similar practical reasons, human beings have long had a need (and, therefore, probably also a desire) to "work multiplication problems backward". For instance, a man who wished to divide his sixty sheep equally among his four sons might wish to know how many sheep each son should receive. Thus he would be faced with the problem of finding a number b such that

$$4 \cdot b = 60.$$

Previous experience with multiplication might enable one to recognize that the number b sought here would have to be 15.

The generalized form of the above problem is: given a product (say c) and one of its factors (say a), to find uniquely the other factor b; that is, given numbers a and c, to find uniquely a number b such that

$$a \cdot b = c$$

or, equivalently (since $a \cdot b = b \cdot a$), such that

$$b \cdot a = c.$$

This type of problem is clearly the inverse of the multiplication problem in which the factors a and b are given and it is required to find a number c such that

$$a \cdot b = c \text{ (and } b \cdot a = c).$$

The process of finding the unique number b (if this can be done) such that

$$a \cdot b = c \text{ (and } b \cdot a = c)$$

is called "**dividing** c by a" and the required number b (regardless of the nature of its value, and regardless of whether this value may be known or not) is called "the **quotient** of c divided by a" or "the **ratio** of c to a". This quotient is represented by the symbol

$$\frac{c}{a}$$

and also by the symbols

$$c/a \text{ and } c \div a,$$

any one of which may be read as "c divided by a", or as "c over a". The numbers c and a are called the **dividend** and the **divisor,** respectively. Notice that we are *defining* the meaning of c/a to be such that the statements

$$a \cdot \frac{c}{a} = c \text{ and } \frac{c}{a} \cdot a = c$$

are true.

Thus, for example,

$$\frac{60}{4} = 15$$

because of the fact that

$$4 \cdot 15 = 60 \text{ (and } 15 \cdot 4 = 60).$$

We note that the truth of the statements

$$4 \cdot \frac{60}{4} = 60 \quad \text{and} \quad \frac{60}{4} \cdot 4 = 60$$

follows automatically from the definition of the meaning of the symbol 60/4, and hence needs no further proof.

1.4.3.1.1. *Divisibility of integers; even and odd integers.* If for some two integers a and c, there can be found a unique integer b such that

$$a \cdot b = c \text{ (and } b \cdot a = c),$$

then we say that c **is exactly divisible by** a and that a is an **exact divisor,** or **factor,** of c.

An integer that is exactly divisible by two is called an **even integer.** Each integer which is not an even integer is called an **odd integer.** Thus, for example, —14 is an even integer because $-14 = -7 \cdot 2$, whereas 13, 15, 17, —17, etc., are all odd integers.

1.4.3.1.2. "*You can't divide by zero*". Recalling that

$$b \cdot 0 = 0 \text{ and } 0 \cdot b = 0$$

are true statements for any integer b, we shall see that this fact makes it impossible for us to consider the number zero as a divisor. To see why, suppose that

$$\frac{c}{0} = b.$$

Then according to our definition of the meaning of a quotient, this means that

$$0 \cdot b = c \text{ (and } b \cdot 0 = c).$$

These statements are clearly false for any number c which is different from zero since the products $0 \cdot b$ and $b \cdot 0$ are both zero for any number b. Thus if c is not zero it is not possible to find any number b such that $c/0 = b$; that is, it is not possible to divide c by 0. In the one remaining case, in which $c = 0$, the situation is no more satisfactory for, altho the statement $0 \cdot b = c$ is true, it is equally true for any and every number which the symbol b might represent, and hence this does not make it possible for us to determine any *unique* (that is, *particular* or *definite*) value of b to recognize as the meaning of the symbol 0/0. Thus **the symbol 0/0 has no definite meaning as a number and the symbol $c/0$ (where c is different from 0) has no meaning at all as a**

number. Accordingly, it is clear that **division by zero is not possible in any case.**

It should be noted here that the foregoing discussion concerns only division *by* zero, and not division *of* zero by another number. Division *of* zero *by* a number (say a) which is different from zero is perfectly possible and for any such non-zero number a it is true that

$$\frac{0}{a} = 0$$

because of the fact that

$$a \cdot 0 = 0 \text{ and } 0 \cdot a = 0.$$

1.4.3.2. *Division and the natural laws*

Since division is a binary operation, we are naturally curious to know which of the natural laws, if any, is valid for this operation. As in the case of subtraction, we can answer most aspects of this question by considering a few examples.

Since there is no integer b such that

$$2 \cdot b = 1$$

it is clear that the quotient 1/2, if there is any such "number" at all, is certainly not an integer. Thus **the system of integers is not closed for division.**

Further, since

$$2/1 = 2$$

(because $1 \cdot 2 = 2$ and $2 \cdot 1 = 2$)
and thus 2/1 is an integer, while 1/2 is not, it is clear that 2/1 and 1/2 are not equal. Thus **division is not commutative.**

If division is to be associative then it must be true that for any three integers a, b and c (with b and c different from zero),

$$(a/b)/c = a/(b/c).$$

To test this, we consider the case in which $a = 24$, $b = 4$ and $c = 2$. For this case

$$(a/b)/c = (24/4)/2 = 6/2 = 3$$

and

$$a/(b/c) = 24/(4/2) = 24/2 = 12.$$

Since $(a/b)/c$ and $a/(b/c)$ are not equal in this case, it is seen that **division is not associative.**

While it is true that $a/1 = a$ for any integer a, there is no integer e such that $e/a = a$ for every integer a. Hence **there is no identity element for division in the system of integers.**

We say that "multiplication is distributive over addition" to describe the fact that for any three integers a, b and c,

$$a \cdot (b + c) = a \cdot b + a \cdot c$$

and

$$(b + c) \cdot a = b \cdot a + c \cdot a.$$

We describe the first of these two relations by saying that multiplication *on the left* is distributive over addition, and we describe the second relation by saying that multiplication *on the right* is distributive over addition. Curiosity leads us to inquire whether either or both of the corresponding statements

$$a/(b + c) = (a/b) + (a/c)$$

and

$$(b + c)/a = (b/a) + (c/a)$$

may be true for all integers a, b and c (provided zero is not permitted as a divisor). We shall see that the first statement is generally false but that the second one is true. To prove that the first statement is not generally true we need only consider a properly selected "counter-example" such as the case in which $a = 24$, $b = 4$ and $c = 2$. For this case,

$$a/(b + c) = 24/(4 + 2) = 24/6 = 4$$

and

$$(a/b) + (a/c) = (24/4) + (24/2) = 6 + 12 = 18.$$

Since the statement is not true in this case, we know at once it is not generally true.

To prove that the second statement is true in all cases (where a is different from zero), we argue as follows. Suppose that

$$b/a = d \text{ and } c/a = e.$$

This means that there are unique integers d and e such that

$$a \cdot d = \text{b and } a \cdot e = c$$

so that

$$a \cdot d + a \cdot e = b + c.$$

But by the distribution law for multiplication over addition,

$$a \cdot d + a \cdot e = a \cdot (d + e),$$

and hence we know that

$$a \cdot (d + e) = b + c.$$

By our definition of the quotient concept, this means that

$$(b + c)/a = d + e,$$

or, recalling that $d = b/a$ and $e = c/a$,

$$(b + c)/a = (b/a) + (c/a).$$

This relation represents the *only* aspect in which division is distributive over addition. In connection with the use of the symbol " / " for division, it might be described by saying that "division **on the right** is distributive over addition". In a similar manner, it can be shown that "division **on the right** is distributive over subtraction", meaning that for any three integers a, b and c (with a different from zero),

$$(b - c)/a = (b/a) - (c/a).$$

The counter-example we considered makes it clear that division **on the left** is **not** distributive over addition and hence that, in general,

$$a/(b + c) \neq (a/b) + (a/c).$$

(The symbol " $\neq$ " is to be read as "is not equal to".) Similarly, division **on the left** is **not** distributive over subtraction so that, in general,

$$a/(b - c) \neq (a/b) - (a/c).$$

1.4.3.2.1. *Explanation of optional use or omission of grouping symbols in certain expressions.* It will be recalled that in discussing the distributive law

$$(a + b) \cdot c = a \cdot c + b \cdot c,$$

the expression to the right of the equality symbol must be interpreted as the sum of the products $a \cdot c$ and $b \cdot c$; that is, as

$$(a \cdot c) + (b \cdot c),$$

even tho we do not ordinarily bother to write the grouping symbols around the individual products. To be honest we must admit that this could be confusing to someone not used to the custom and that it would be excusable for such a person to wonder if, for instance, the expressions $a \cdot c + b \cdot c$ and $a \cdot (c + b) \cdot c$ might also be equivalent and interchangeable, or if $(a + b) \cdot c$ and $a + b \cdot c$ might be equivalent and interchangeable, etc. The answer to each of the latter two questions is a definite "No!" and the explanation of this seemingly inconsistent state of affairs is the following.

Purely as a matter of custom it has been agreed to recognize that multiplication has "precedence" over addition. This means that in any symbolic representation of a computation in which both of these operations are used and no grouping symbols appear, the multiplication operations are always understood to be performed first and the addition operations are then performed on the results of the multiplications. Thus, for example, we understand (by agreement) that $2 \cdot 3 + 4 \cdot 5$ means $(2 \cdot 3) + (4 \cdot 5)$ or $6 + 20$ or, finally, 26. Similarly, the expression $2 + 3 \cdot 4$, written without grouping symbols, means $2 + (3 \cdot 4)$ or $2 + 12$, or 14.

In any case where it is desired to perform one or more additions first and then perform one or more multiplications on the results of these additions, this *must* be indicated by the use of grouping symbols written around the indicated sums. Thus, for example,

$$2 \cdot (3 + 4) \cdot 5 = 2 \cdot 7 \cdot 5 = 70$$
$$(2 + 3) \cdot 4 = 5 \cdot 4 = 20.$$

In a similar manner, multiplication has precedence over subtraction, so that

$$4 \cdot 3 - 2 = (4 \cdot 3) - 2 = 12 - 2 = 10$$

altho

$$4 \cdot (3 - 2) = 4 \cdot 1 = 4.$$

Likewise, division has precedence over both addition and subtraction.

EXAMPLES

$$12/4 + 2 = (12/4) + 2 = 3 + 2 = 5$$
$$12/(4 + 2) = 12/6 = 2$$

$$12/4 - 2 = (12/4) - 2 = 3 - 2 = 1$$
$$12/(4 - 2) = 12/2 = 6$$

$$18 + 6/3 = 18 + (6/3) = 18 + 2 = 20$$
$$(18 + 6)/3 = 24/3 = 8$$

$$18 - 6/3 = 18 - 2 = 16$$
$$(18 - 6)/3 = 12/3 = 4.$$

1.4.3.3. *Enlargement of the number system to provide closure for division; the rational numbers*

In the case of division, as in that of subtraction, there is not much that we can do to remedy the lack of commutativity, associativity and an identity element. The lack of closure for subtraction, however, was overcome by moving from the natural numbers to the larger system of the integers. Having found that certain differences $a - c$ (such as $3 - 5$, for instance) were not in the natural number system, we proceeded to kick the walls out and enlarge the house to make it accommodate *all differences*. It was found that this could be done by taking in zero and, for each natural number n, a new number $-n$ having the property that $-n + n = 0$ and $n + (-n) = 0$. These new numbers were called negative numbers.

With this precedent to guide us, it doesn't take too much imagination to see what should be done about the lack of closure for division. Having noticed that there are certain ratios (or quotients) c/a, such as 1/2, 3/5, 3/2, etc., which are not in the system of integers (that is, there is no integer b such that $2 \cdot b = 1$, or such that $5 \cdot b = 3$, or such that $2 \cdot b = 3$, etc.), we decide to remedy the situation by inventing a new system of numbers in which these ratios—and all ratios of integers of any sort (as long as the divisor is not zero)—shall be included and recognized as "numbers". We call these new numbers **rational numbers,** naturally. (*Ratio*-nal. See?)

Formally, **for any integers c and a with a different from zero, we define the rational number c/a to be that unique number** (whether it turns out to be an integer or not) **for which**

$$a \cdot (c/a) = c \text{ and } (c/a) \cdot a = c.$$

The integers c and a, as used here, are still called the **dividend** and **divisor** respectively, to be consistent with our earlier definition of a quotient.

We note that any integer a is the rational number $a/1$ since it is true that $1 \cdot a = a$ and $a \cdot 1 = a$.

To see how this definition might reflect the real facts of the physical world, consider the problem of the man who owns a four-mile stretch of clam-loaded beach and wishes to divide it equally among his three sons.

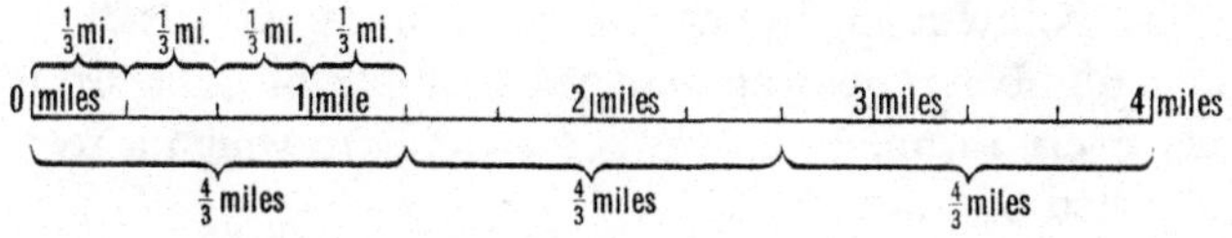

The length of the beach to be given to each son must then be represented by a number b (of miles) which, when taken three times, gives a total length of four miles. That is, the number b is such that

$$b + b + b = 4$$

or

$$3 \cdot b = 4.$$

Hence, according to our definition of the meaning of the rational number 4/3, it is 4/3 miles of beach that each son gets. (The number 4/3 is read as "four-thirds".)

In exactly the same way, if one mile of beach were to be divided into three equal parts, the length of each part (in miles) would be represented by the rational number 1/3 (read as "one-third"). We notice that if such a distance is taken four times a total distance of 4/3 miles is attained. That is,

$$(1/3) + (1/3) + (1/3) + (1/3) = 4/3$$

or

$$4 \cdot (1/3) = 4/3.$$

This illustrates the general principle (which we could verify in the same manner for any case) that for any two integers c and a with a different from zero, the product obtained by multiplying the rational number $1/a$ by the integer c is the rational number c/a, or

$$c \cdot (1/a) = c/a.$$

To provide for the extended applicability of the commutative law for multiplication, we *define* $(1/a) \cdot c$ to have the same meaning as $c \cdot (1/a)$, namely c/a.

A rational number of the form $1/a$, in which the dividend is the number one, is called the **reciprocal** of the number a. What we have just pointed out in the preceding paragraph is that every rational number is just an integral multiple of some reciprocal (that is, every rational number is a product in which a reciprocal is multiplied by an integer).

Referring back to the problem of dividing the four-mile stretch of beach into three equal parts, we notice that the resulting number 4/3 has a divisor 3 whose reciprocal 1/3 *designates* or *denominates* the size of the parts into which each individual mile is divided, and that the dividend 4 *numbers* or *numerates* the times which such a part is taken. For this reason the divisor in such cases is also called the **denominator** and the dividend is called the **numerator**. Because rational numbers are used in this manner to describe **parts** of things, which we may think of as *broken*, or *fractured*, pieces, they are also called **fractions**.

A rather important thing to know about rational numbers is that the same rational number can be written in many different ways. Suppose, for instance, that the clam-baron of our problem had partitioned each mile of his beach into *twice* as many equal parts as before (that is, six rather than three) and had then given each son *twice* as many parts (sixth parts, of course, rather than third parts). It is clear that each son would still receive the same amount of beach as before altho it would now be described by the rational number 8/6 instead of 4/3. Thus, if we double both the denominator (the number of equal parts into which something is partitioned) and the numerator (the number of these equal parts which are to be taken) then the total amount taken is not changed. In the equality

$$\frac{8}{6} = \frac{4}{3}$$

we note that $8 = 2 \cdot 4$ and $6 = 2 \cdot 3$ so that the equality may also be written as

$$\frac{2 \cdot 4}{2 \cdot 3} = \frac{4}{3}.$$

This illustrates the general principle that **for any integers a, b and c with a and c both different from zero, the rational number $(a \cdot b)/(a \cdot c)$ is the same as b/c, or**

$$\frac{a \cdot b}{a \cdot c} = \frac{b}{c}.$$

Since b/c is a simpler form than $(a \cdot b)/(a \cdot c)$, we regard a change from $(a \cdot b)/(a \cdot c)$ to b/c as a *simplification*, and we usually prefer to take advantage of all such simplifications that are possible, unless there is some special reason for preferring a less simple form. When a fraction (or rational number) has been simplified as much as possible from this point of view, it is then said to be "in its **lowest terms**" and the process of obtaining such simplification is called "**reducing** the fraction to its lowest terms".

1.4.3.3.1. *Operations with the rational numbers.* Now that we know what rational numbers are, the next thing to learn about them is how to perform the various binary operations with them.

Suppose we wish to add the two rational numbers a/b and c/b, where a, b and c are integers and b is not zero. Let us suppose that

$$\frac{a}{b} = d \text{ and } \frac{c}{b} = e,$$

where we are using the single letters d and e to represent the *single rational numbers* determined by the *pairs* of *integers* (a, b) and (c, b) respectively. (Do not make the mistake of assuming that the symbol d represents an integer merely because it is only one letter. Letters may represent anything at all, and in any particular case they represent only what we agree that they shall represent. Here we are agreeing that the symbol d shall represent the rational number a/b and that the symbol e shall represent the rational number c/b.) Then, by our definition of rational numbers, this, of course, means that

$$b \cdot d = a \text{ and } b \cdot e = c$$

whence

$$b \cdot d + b \cdot e = a + c,$$

or, using the distributive law of multiplication over addition,

$$b \cdot (d + e) = a + c.$$

Again referring to the definition of rational numbers, we see that this statement is equivalent to

$$\frac{a + c}{b} = d + e.$$

Now replacing the symbols d and e by what they represent, we have

$$\frac{a+c}{b} = \frac{a}{b} + \frac{c}{b}$$

or

$$\frac{a}{b} + \frac{c}{b} = \frac{a+c}{b}.$$

It is worth noting that this rule for adding rational numbers (which have the same denominator) is exactly the same as the statement of the distributivity of division over addition. In the same way, the statement of the distributivity of division over subtraction,

$$\frac{a-c}{b} = \frac{a}{b} - \frac{c}{b}$$

when written backward, becomes our rule

$$\frac{a}{b} - \frac{c}{b} = \frac{a-c}{b}$$

for the subtraction of rational numbers which have the same denominator.

If we wish to add two fractions, say a/b and c/d, which have *different* denominators, we must first change the form of the fractions (by using the reduction principle that $(a \cdot b)/(a \cdot c) = (b/c)$ in such a way that they are made to appear, after the change, with the *same* denominator. Thus, since we can write a/b as $(d \cdot a)/(d \cdot b)$ or $(a \cdot d)/(b \cdot d)$, and c/d can be written as $(b \cdot c)/(b \cdot d)$ we have

$$\frac{a}{b} + \frac{c}{d} = \frac{a \cdot d}{b \cdot d} + \frac{b \cdot c}{b \cdot d}$$

$$= \frac{a \cdot d + b \cdot c}{b \cdot d}.$$

Similarly,

$$\frac{a}{b} - \frac{c}{d} = \frac{a \cdot d - b \cdot c}{b \cdot d}.$$

[Notice well that in writing a/b as $(a \cdot d)/(b \cdot d)$, we have merely written the *same* rational number in a different way, that is, we have changed the *appearance* of the symbol representing the number but *the value of the number has not been changed,* and the two symbols a/b and $(a \cdot d)/(b \cdot d)$ both describe exactly the same rational number.]

As a matter of notation, the multiplication symbol $\cdot$ is frequently omitted in cases where no confusion, ambiguity or misunderstanding of any sort will result from such omission. We may do this frequently from now on as a matter of convenience. Thus the above results may be written in the form

$$\frac{a}{b} + \frac{c}{d} = \frac{ad + bc}{bd},$$

$$\frac{a}{b} - \frac{c}{d} = \frac{ad - bc}{bd}.$$

To find the product of two rational numbers, say a/b and c/d, suppose that

$$\frac{a}{b} = e \text{ and } \frac{c}{d} = f.$$

This means that e and f are rational numbers such that

$$be = a \text{ and } df = c$$

and hence that

$$(be)(df) = ac.$$

If we assume, as would be convenient and desirable, that multiplication of rational numbers is both commutative and associative, then the product $(be)(df)$ can be rewritten as $(bd)(ef)$ and it follows that

$$(bd)(ef) = ac$$

which, according to the definition for rational numbers, means that

$$\frac{ac}{bd} = ef, \text{ or } \frac{a \cdot c}{b \cdot d} = e \cdot f.$$

But we also have

$$\frac{a}{b} \cdot \frac{c}{d} = e \cdot f.$$

Hence

$$\frac{a}{b} \cdot \frac{c}{d} = \frac{a \cdot c}{b \cdot d}.$$

In words, the rule for finding the product of two fractions is: *the product of two fractions is the product of their numerators divided by the product of their denominators.*

Since the preceding result was obtained as a consequence of our assumption of the commutative and associative laws for multiplication of rational numbers, it would be equally correct to say that we *defined* the product $(a/b) \cdot (c/d)$ to be $(a \cdot c)/(b \cdot d)$, and did so *in order that* multiplication in the rational number system would be consistent with the commutative and associative laws.

Since any integer a is also the rational number $a/1$, we can use the preceding product formula to find a special rule for multiplying a fraction by an integer. Thus, for any integer a and any fraction b/c with c different from zero, we have

$$a \cdot \frac{b}{c} = \frac{a}{1} \cdot \frac{b}{c} = \frac{a \cdot b}{1 \cdot c} = \frac{ab}{c}.$$

By assuming commutative multiplication for the rational numbers it would then also be true that

$$\frac{b}{c} \cdot a = \frac{ab}{c}.$$

In considering how one fraction should be divided by another we realize that such a quotient is a new concept and calls for a new definition. We decide to make our definition consistent with previous definitions for quotients (of other kinds) and define the quotient

$$\frac{\frac{a}{b}}{\frac{c}{d}}$$

to be that rational number e/f such that

$$\frac{c}{d} \cdot \frac{e}{f} = \frac{a}{b},$$

or

$$\frac{ce}{df} = \frac{a}{b}.$$

It can be seen that if $e = ad$ and $f = bc$, then

$$\frac{ce}{df} = \frac{c(ad)}{d(bc)} = \frac{c(da)}{d(cb)} = \frac{(cd)a}{(dc)b} = \frac{(cd)\,a}{(cd)b} = \frac{a}{b}.$$

Thus if

$$\frac{e}{f} = \frac{ad}{bc},$$

then

$$\frac{c}{d} \cdot \frac{e}{f} = \frac{a}{b}$$

and hence

$$\frac{\frac{a}{b}}{\frac{c}{d}} = \frac{ad}{bc},$$

or, alternatively,

$$\frac{\frac{a}{b}}{\frac{c}{d}} = \frac{a}{b} \cdot \frac{d}{c}.$$

A useful additional consequence of this result is found thru its application to the simplification of the reciprocal of the fraction c/d by considering the case in which a/b is the rational number one. We get

$$\frac{1}{\frac{c}{d}} = 1 \cdot \frac{d}{c} = \frac{1 \cdot d}{c} = \frac{d}{c},$$

or, in words, **the fraction d/c is the reciprocal of the fraction c/d.**

We have now developed a system of numbers, the rational numbers, in which we have a closure law for each of the binary operations of addition, multiplication, subtraction and division and commutative and associative laws for both addition and multiplication, in which multiplication is distributive over both addition and subtraction, and in which division is, in one sense, distributive over both addition and subtraction. This is a very satisfactory number system for many purposes and history records that it served mankind very well for many centuries.

1.4.3.4. *Division considered as extended subtraction*

Altho we have discussed the operation of division as the inverse of multiplication it is also possible, useful and instructive to consider it as extended subtraction.

The definition we have used provides that, for any positive integers a and c with a different from zero,

$$\frac{c}{a} = b$$

if b is a unique number for which

$$ab = c \text{ (and } ba = c).$$

If b is a positive integer, then we may interpret the meaning of the quotient b as either: that b is the number which must be taken exactly a times (in addition) to produce c, or that b is the precise number of times which the number a must be taken (in addition) to produce the number c. Taking the latter viewpoint that it takes b as to produce c, then it is clear that the number a could be taken from c once, twice, or any number of times up to and including b times. If a is taken from c fewer than b times, then some positive difference will remain; but if a is taken from c exactly b times, then all of a will have been exhausted and only zero will remain. That is, if $c = b \cdot a$, then

$$c - a = (b - 1) \cdot a$$
$$c - a - a = (b - 2) \cdot a, \text{ or } c - 2a = (b - 2) \cdot a$$
$$\cdot \quad \cdot \quad \cdot$$
$$c - b \cdot a = 0,$$

and the integer b is seen as the maximum number of times which a may be taken from c before c is exhausted.

On the other hand, if the quotient c/a is not a positive integer but some non-integral positive rational number, then after taking a from c the maximum possible positive integral number of times, say b, there will still remain some integer r which is some one of the numbers 1, 2, 3, ..., $a - 1$. That is,

$$c - b \cdot a = r$$

or

$$c = b \cdot a + r$$

and hence

$$\frac{c}{a} = \frac{b \cdot a + r}{a} = \frac{b \cdot a}{a} + \frac{r}{a} = b + \frac{r}{a}.$$

Thus the quotient c/a, in such a case as this, is the sum of an integer b and a fraction r/a in which the numerator is less than the denominator. Since the integer r is what *remains* after the number a has been taken from c the maximum possible number b of times, it is clear that r is what is properly to be designated as the **remainder** in such a division problem. While the fraction r/a is sometimes mistakenly called the remainder and the integer b is quite commonly called the quotient, if we speak correctly it is c/a which must be called "the quotient" while b is the **integral part** of the quotient and r/a is the **fractional part** of the quotient.

Exercises

1. State in one sentence (not to exceed one line in length) *why* 78/6 is 13.

2. (a) What is the reciprocal of 8? (b) What is the reciprocal of 1/5? (c) What is the reciprocal of the reciprocal of 17? (d) What is the reciprocal of the reciprocal of the number n? Is there any exception to the answer you have given? (e) What is the reciprocal of -1?

3. From the definition of a quotient and the known facts of multiplication in the system of integers deduce the appropriate conclusion (positive or negative) for each of the following statements.

(a) The quotient of a positive number divided by a positive number is . . . .
(b) The quotient of a positive number divided by a negative number is . . . .
(c) The quotient of a negative number divided by a positive number is . . . .
(d) The quotient of a negative number divided by a negative number is . . . .

4. (a) Knowing that $10^2 = 100$, $10^3 = 1000$, $10^4 = 10{,}000$, etc., if we are to *define* values for 10^1, 10^0, 10^{-1}, 10^{-2}, 10^{-3} and 10^{-4}, what values would you recommend as seeming most appropriate? (b) For any positive integer n, what would you consider to be a proper meaning to assign (as a matter of definition) to the symbol 10^{-n}?

5. Discuss the matter of whether or not the operation of division has an inverse.

6. What is the value of the fraction $(x - y)/(y - z)$ if (a) $x = 4$, $y = 4$, $z = 2$? (b) $x = 4$, $y = 2$, $z = 4$? (c) $x = 2$, $y = 4$, $z = 4$? (d) $x = 4$, $y = 4$, $z = 4$?

7. Reduce each of the following fractions to its lowest terms. Explain any exceptional case.

(a) $\frac{14}{2}$ (b) $\frac{195}{65}$ (c) $\frac{784}{48}$ (d) $\frac{7}{0}$ (e) $\frac{0}{11}$ (f) $\frac{72}{48}$

(g) $\frac{432}{172}$ (h) $\frac{756}{42}$ (i) $\frac{29}{3}$ (j) $\frac{77}{42}$.

8. Express each of the following quotients formally as the sum of its integral and fractional parts. (Reduce the fractional part to lowest terms.) Explain any exceptional case.

(a) $\frac{17}{5}$ (b) $\frac{15}{3}$ (c) $\frac{132}{24}$ (d) $\frac{1}{7}$ (e) $\frac{13}{0}$.

9. Express each of the following numbers formally as the ratio of two integers in lowest terms.

(a) $14\frac{3}{7}$ (b) 18 (c) $10\frac{4}{10}$ (d) 0 (e) $493\frac{17}{59}$.

10. Find the integral quotient and the remainder in each of the following division problems and write the formal relation which connects dividend, divisor, quotient and remainder. (EXAMPLE: For 17/7, the integral quotient is 2 and the remainder is 3. The required relation is $17 = 2 \cdot 7 + 3$.)
(a) 49/11 (b) 2612/43 (c) 4716/18 (d) 41/139 (e) 217/79.

11. All of the following are rational numbers, some being integers and some not. List the ones which are integers.

(a) −18 (b) 2/6 (c) 6/2 (d) −25/100 (e) 0/13 (f) 22/7
(g) 195,000/100 (h) −100/25 (i) 1 (j) $33\frac{1}{3}$.

12. Reduce each of the following expressions to its simplest numerical form. Explain any exceptional case.

(a) $\frac{24 - 8}{6 - 4}$ (b) $\frac{12 - 4}{2 - 6}$ (c) $\frac{42 - 42}{26 - 5}$ (d) $\frac{54 - 45}{9 - 9}$ (e) $\frac{6 - 6}{6 - 6}$

(f) $\frac{4 - 24}{8 - 12}$ (g) $\frac{-216}{6 - 9}$.

13. Explain why, in defining the concept of "inverse multiplication" (that is, division) on page 82, we insist that the required number b be unique.

14. Draw a diagram to represent 3/5 of a line segment 7 inches long.

15. Draw a rectangle and then show how to represent 4/7 of its area.

16. Draw a circle and then show how to represent 5/8 of its area.

17. Use a rectangular diagram to demonstrate why $(2/3) \cdot (5/7)$ is $(2 \cdot 5)/(3 \cdot 7)$.

18. If a line segment of a specified length is given, can a rational number be found to represent the length of any specified part of it?

19. (a) List the first fifteen positive even integers. (b) List the first fifteen positive odd integers. (c) List the first fifteen prime numbers. (d) List the first fifteen odd prime numbers. (e) What even prime numbers are there?

20. List all the prime numbers among the first hundred positive integers.

21. Is the set of non-integral rational numbers closed under (a) addition? (b) multiplication? (c) subtraction? (d) division?

1.4.3.5. *Prime numbers*

A **prime** number is a positive integer, greater than one, which has only itself and the number one as factors. Examples are 2, 3, 5, 7, 11, 13, 17, 19, 23, 29, etc. A positive integer which is not a prime is called a **composite** number.

1.4.4. Convergence; the Real Numbers

As the failure of the system of counting numbers to provide closure for subtraction led to the development of the system of integers and the failure of this latter system to provide closure for division led to the development of the system of rational numbers, in the same manner the rational numbers are found not to be closed for the operations which are inverse to exponentiation and this leads to the development of a still more extensive number system.

1.4.4.1. *Extension of the definition of exponentiation*

In order to appreciate the circumstances motivating this further expansion of the number system, let's examine the properties of exponentiation and its inverses in the rational number system.

The general problem of exponentiation is this: Given two numbers a and b, find a number c such that

$$a^b = c,$$

where the power a^b is defined to mean a product of b factors, each equal to a. Of course, such a definition has obvious meaning only if b is a positive integer equal to or greater than two. Thus

$$\begin{aligned} a^2 &= a \cdot a \\ a^3 &= a \cdot a \cdot a \\ a^4 &= a \cdot a \cdot a \cdot a, \text{ etc.} \end{aligned}$$

From the definition stated above it is not entirely clear what the meaning of a^1 should be since a is not ordinarily considered as a product (unless we consider it as the product $1 \cdot a$) and it is not immediately evident what should be the meaning (if any) of "powers" such as a^0, a^{-1}, a^{-2}, $a^{1/2}$, $a^{1/3}$, $a^{5/8}$, etc. To clear up some of these questions let us observe the behavior of certain powers under the operations of multiplication, division and exponentiation.

By referring to our definition of a power, the product of two powers with the same base, such as a^b and a^c, can be found as follows.

$$\begin{aligned} a^b \cdot a^c &= \overbrace{a \cdot a \cdot a \cdot \, \ldots \, \cdot a}^{b \text{ factors}} \cdot \overbrace{a \cdot a \cdot a \cdot \, \ldots \, \cdot a}^{c \text{ factors}} \\ &= \overbrace{a \cdot a \cdot a \cdot \qquad \ldots \qquad \cdot a}^{(b+c) \text{ factors}} \\ &= a^{b+c}. \end{aligned}$$

Similarly, if b is greater than c, then

$$\frac{a^b}{a^c} = \frac{\overbrace{a \cdot a \cdot a \cdot \quad \ldots \quad \cdot a}^{b \text{ factors}}}{\underbrace{a \cdot a \cdot a \cdot \, \ldots \, \cdot a}_{c \text{ factors}}} = \frac{\overbrace{a \cdot a \cdot a \cdot \, \ldots \, \cdot a}^{c \text{ factors}} \cdot \overbrace{a \cdot a \cdot a \cdot \, \ldots \, \cdot a}^{(b-c) \text{ factors}}}{\underbrace{a \cdot a \cdot a \cdot \, \ldots \, \cdot a}_{c \text{ factors}}}.$$

By the principle discovered in section 1.4.3.3 for reducing fractions to lowest terms, it then follows that

$$\frac{a^b}{a^c} = \overbrace{a \cdot a \cdot a \cdot \ \ldots \ \cdot a}^{(b-c) \text{ factors}} = a^{b-c}.$$

Also

$$\begin{aligned}(a^b)^c &= (\overbrace{a \cdot a \cdot a \cdot \ \ldots \ \cdot a}^{b \text{ factors}})^c \\ &= \underbrace{(\overbrace{a \cdot a \cdot a \cdot \ \ldots \ \cdot a}^{b \text{ factors}}) \cdot (\overbrace{a \cdot a \cdot a \cdot \ \ldots \ \cdot a}^{b \text{ factors}}) \cdot \ \ldots \ \cdot (\overbrace{a \cdot a \cdot a \cdot \ \ldots \ \cdot a}^{b \text{ factors}})}_{c \text{ sets of } b \text{ factors each}} \\ &= \overbrace{a \cdot a \cdot a \cdot \qquad \ldots \qquad \cdot a}^{bc \text{ factors}} \\ &= a^{bc}.\end{aligned}$$

These three principles,

$$\boldsymbol{a^b \cdot a^c = a^{b+c}}$$

$$\boldsymbol{\frac{a^b}{a^c} = a^{b-c}} \text{ **(if } b \text{ is greater than } c\text{)**}$$

$$\boldsymbol{(a^b)^c = a^{bc}},$$

are of fundamental importance in working with powers and are known as "**laws of exponents**". The following examples will illustrate their use.

$$a^3 \cdot a^4 = (a \cdot a \cdot a) \cdot (a \cdot a \cdot a \cdot a) = a \cdot a \cdot a \cdot a \cdot a \cdot a \cdot a = a^7 = a^{3+4}$$

$$\begin{aligned}\frac{a^8}{a^5} &= \frac{a \cdot a \cdot a \cdot a \cdot a \cdot a \cdot a \cdot a}{a \cdot a \cdot a \cdot a \cdot a} = \frac{(a \cdot a \cdot a \cdot a \cdot a) \cdot (a \cdot a \cdot a)}{a \cdot a \cdot a \cdot a \cdot a} = \frac{a^5 \cdot a^3}{a^5} \\ &= a^3 = a^{8-5}\end{aligned}$$

$$\begin{aligned}(a^3)^4 &= (a \cdot a \cdot a)^4 = (a \cdot a \cdot a) \cdot (a \cdot a \cdot a) \cdot (a \cdot a \cdot a) \cdot (a \cdot a \cdot a) \\ &= a \cdot a \cdot a \cdot a \cdot a \cdot a \cdot a \cdot a \cdot a \cdot a \cdot a \cdot a = a^{12} = a^{3 \cdot 4}\end{aligned}$$

or, from an alternative viewpoint,

$$(a^3)^4 = a^3 \cdot a^3 \cdot a^3 \cdot a^3 = a^{3+3+3+3} = a^{3 \cdot 4}.$$

It may sometimes happen, of course, that in using the laws of exponents, someone may become more enthusiastic than cautious and use one of the laws in a case for which it was not designed. For instance, if one forgets that the law

$$\frac{a^b}{a^c} = a^{b-c}$$

was designed to be used only when b is greater than c, and uses it in an attempt to simplify the quotient a^5/a^8, he may appear to find that

$$\frac{a^5}{a^8} = a^{5-8} = a^{-3}.$$

Actually, of course,

$$\frac{a^5}{a^8} = \frac{a \cdot a \cdot a \cdot a \cdot a}{a \cdot a \cdot a \cdot a \cdot a \cdot a \cdot a \cdot a} = \frac{a \cdot a \cdot a \cdot a \cdot a}{(a \cdot a \cdot a \cdot a \cdot a) \cdot (a \cdot a \cdot a)} = \frac{a^5}{a^5 \cdot a^3} = \frac{1}{a^3}.$$

Thus we are reminded that we should not have used the law

$$\frac{a^b}{a^c} = a^{b-c}$$

for this case—that is, unless we are prepared to recognize a^{-3} as meaning $1/a^3$. If we think about this for a bit, we may realize that this is exactly the clue we have been looking for to help us decide what meaning should be associated with a^{-3}. Taking this as our cue, we extend our definition of powers by *defining* that, **for any positive integer n, a^{-n} shall mean $1/a^n$.** (It should be understood that our purpose here is to take advantage of our prerogative of extending the definition of exponentiation by doing it in such a way as to make the law

$$\frac{a^b}{a^c} = a^{b-c}$$

consistently valid in all cases.)

In a similar manner, since the use of this same law yields

$$\frac{a^b}{a^b} = a^{b-b} = a^0$$

whereas we know from other considerations that $a^b/a^b = 1$, **we define $a^0 = 1$.**

Likewise, since the same law yields

$$\frac{a^{c+1}}{a^c} = a^{(c+1)-c} = a^1,$$

but we know that

$$\frac{a^{c+1}}{a^c} = \frac{\overbrace{a \cdot a \cdot a \cdot \ \dots \ \cdot a}^{(c+1) \text{ factors}}}{\underbrace{a \cdot a \cdot a \cdot \ \dots \ \cdot a}_{c \text{ factors}}} = \frac{\overbrace{(a \cdot a \cdot a \cdot \ \dots \ \cdot a)}^{c \text{ factors}} \cdot a}{\underbrace{a \cdot a \cdot a \cdot \ \dots \ \cdot a}_{c \text{ factors}}} = a,$$

we define $a^1 = a$.

We now have the following consistent pattern of definitions for exponentiation with integral exponents:

$$\begin{aligned}
&\cdot \quad \cdot \quad \cdot \\
a^{-4} &= 1/(a \cdot a \cdot a \cdot a) \\
a^{-3} &= 1/(a \cdot a \cdot a) \\
a^{-2} &= 1/(a \cdot a) \\
a^{-1} &= 1/a \\
a^0 &= 1 \\
a^1 &= a \\
a^2 &= a \cdot a \\
a^3 &= a \cdot a \cdot a \\
a^4 &= a \cdot a \cdot a \cdot a \\
&\cdot \quad \cdot \quad \cdot \cdot
\end{aligned}$$

In determining an appropriate meaning for such expressions as $a^{1/2}$, $a^{1/3}$, $a^{5/8}$, etc., we allow ourselves to be guided by the principle of the law $(a^b)^c = a^{bc}$. Thus we let the nature of $a^{1/2}$ be defined by the relation

$$(a^{1/2})^2 = a^{(1/2)\cdot 2} = a^1 = a$$

to be a number whose second power is a. Such a number is called a "**second root** of a" (or, more commonly, a **square root** of a, since it is customary to refer to the second power of a number as its "square"). For example,

$$4^{1/2} = 2 \text{ since } 2^2 = 4,$$
$$25^{1/2} = 5 \text{ since } 5^2 = 25,$$
$$(1/9)^{1/2} = 1/3 \text{ since } (1/3)^2 = 1/9, \text{ etc.}$$

Similarly, we let $a^{1/3}$ be defined by the relation

$$(a^{1/3})^3 = a^{(1/3)\cdot 3} = a^1 = a$$

as a number whose third power is a. Such a number is called a "**third root** of a" (or, also, a **cube root** of a, since third powers are called "cubes"). For example,

$$8^{1/3} = 2 \text{ since } 2^3 = 8,$$
$$(.001)^{1/3} = .1 \text{ since } (.1)^3 = .001, \text{ etc.}$$

In general, we define $a^{1/n}$ by the relation

$$(a^{1/n})^n = a;$$

that is, $a^{1/n}$ is defined as a number whose nth power is a. Hence we say

$$a^{1/n} = b$$

and call b an ***n*th root** of a if

$$b^n = a.$$

If a number has two (or more) nth roots and one of these is a positive number then this positive root is called the **principal *n*th root** of the original number. For example, 2 and -2 are both second roots (or square roots) of 4 since

$$2^2 = 4 \text{ and also } (-2)^2 = 4.$$

Here 2 is the principal second root of 4 and we write

$$4^{1/2} = 2$$

and

$$-4^{1/2} = -2.$$

[Notice that $-4^{1/2}$ means $-(4^{1/2})$, *not* $(-4)^{1/2}$. If we wished to discuss the second root of -4, we would have to represent it symbolically as $(-4)^{1/2}$, using the parentheses to "group" the negative sign with the numeral 4.]

In a similar manner **we define** $a^{5/8}$ by the relation

$$(a^{5/8})^8 = a^{(5/8)\cdot 8} = a^5$$

to be a number whose eighth power is a^5. Since, however,

$$5/8 = 5 \cdot (1/8) = (1/8) \cdot 5,$$

we also have

$$a^{5/8} = a^{5\cdot(1/8)} = a^{(1/8)\cdot 5}$$

and *if* we are to be guided *consistently* by the principle implied in the law $(a^b)^c = a^{bc}$, then we must also interpret $a^{5/8}$ as both $(a^5)^{1/8}$ and $(a^{1/8})^5$; that is, **we must interpret $a^{5/8}$ as both an eighth root of a^5 and the fifth power of $a^{1/8}$.** For example, $256^{5/8}$ is to be interpreted in the following three ways (the second being merely a rephrasing of the first):

(1) as a positive number whose eighth power is the fifth power of 256;
(2) as the principal eighth root of the fifth power of 256;
(3) as the fifth power of the principal eighth root of 256.

The third interpretation is the easiest one to evaluate in this case since 2 is the principal eighth root of 256 (because $2^8 = 256$) and the fifth power of 2 is 32. Thus

$$256^{5/8} = 32.$$

The first two interpretations are seen to apply also since

$$256^5 = 1{,}099{,}511{,}627{,}776$$

and

$$(1{,}099{,}511{,}627{,}776)^{1/8} = 32 \quad (\text{because } 32^8 = 1{,}099{,}511{,}627{,}776).$$

(Notice how much simpler it is to work with the third interpretation even tho the others are equally correct.)

In general, for any integers m and n (with n positive), we shall interpret the symbol $a^{m/n}$ to mean a number whose nth power is a^m, and hence also either **the principal nth root of a^m** or **the mth power of the principal nth root of a.**

This completes our definition of exponentiation for any rational number used as a base and any rational number used as an exponent (with these exceptions: we have not defined 0^0 and we shall not attempt to do so since it is a numerically meaningless symbol, like 0/0; also no numerical meaning will be defined for 0^{-n} for any positive rational number n).
(Note: The archaic symbol $\sqrt[n]{a}$ is still commonly used to mean $a^{1/n}$.)

1.4.4.2. *Exponentiation and the natural laws.*

While exponentiation with any non-zero rational number as a base and any integer as an exponent yields a rational number (as does also exponentiation with zero as a base and any positive rational number as exponent), we find that when we construct a power with a rational base and an arbitrary rational number as exponent, the resulting power is not always a rational number. As an example,

$$2^{1/2}$$

is not a rational number since (as will be shown later) there is no rational number a such that $a^2 = 2$. Thus **the rational numbers are not closed under exponentiation.**

It can be demonstrated quickly that **exponentiation is not commutative** since, for instance, $2^3 = 8$, $3^2 = 9$, and hence

$$2^3 \neq 3^2.$$

Similarly, since

$$(4^3)^2 = 64^2 = 4096$$

and

$$4^{(3^2)} = 4^9 = 262{,}144,$$

it is seen that

$$(4^3)^2 \neq 4^{(3^2)}$$

and hence **exponentiation is not associative.**

To examine exponentiation for distributive properties it is helpful to replace the conventional notation a^b by the more symmetric notation $a * b$.

We shall prove that **exponentiation is distributive over multiplication in one way but that in another way it is not.**

$$(a \cdot b) * c = (a \cdot b)^c = \overbrace{(a \cdot b) \cdot (a \cdot b) \cdot (a \cdot b) \cdot \ldots \cdot (a \cdot b)}^{c \text{ factors, each equal to } (a \cdot b)}$$

$$= \overbrace{(a \cdot a \cdot a \cdot \ldots \cdot a)}^{c \text{ factors}} \cdot \overbrace{(b \cdot b \cdot b \cdot \ldots \cdot b)}^{c \text{ factors}}$$

$$= a^c \cdot b^c$$

$$= (a * c) \cdot (b * c).$$

Thus exponentiation on the right is distributive over multiplication. However,

$$c * (a \cdot b) = c^{ab}$$

and

$$(c * a) \cdot (c * b) = c^a \cdot c^b = c^{a+b},$$

and since ab and $a + b$ are generally not equal, it follows that neither are c^{ab} and c^{a+b}, and hence that, in general,

$$c * (a \cdot b) \neq (c * b) \cdot (c * b).$$

Thus **exponentiation on the left is not distributive over multiplication.**

The one valid distributive principle which we have proved here, and which is usually written in the form

$$(a \cdot b)^c = a^c \cdot b^c,$$

is usually listed as a **fourth law of exponents** along with the three laws of exponents we have already discussed on page 102.

In a similar manner it can be shown that there is no generally valid distributive principle involving multiplication over exponentiation, exponentiation over addition, or addition over exponentiation.

Also, **there is no identity element for exponentiation** since, altho it is true that for every rational number a

$$a * 1 = a^1 = a,$$

it is *not true* that for every rational number a

$$1 * a = 1^a = a$$

because, actually, $1^a = 1$ for any number a.

1.4.4.3. *The operations which are inverse to exponentiation*

The two inverse exponentiation problems are:

(1) given two numbers b and c, to find a number a such that $a^b = c$;

(2) given two numbers a and c, to find a number b such that $a^b = c$.

The inverse operation by which the first problem is solved is seen to be the process of finding the bth root of c. This operation is called **root extraction.** The inverse operation by which the second problem is solved is called **anti-exponentiation** and the number b which is found in this case is called "the **logarithm** of c to the base a" and is written as

$$\log_a c.$$

EXAMPLES

1. Given $b = 5$ and $c = 32$, it is seen that the number a such that $a^5 = 32$ is uniquely determined as 2. Thus 2 is the fifth root of 32.

2. Given $a = 10$ and $c = 1000$, it is seen that the number b such that $10^b = 1000$ is uniquely determined as 3. Thus 3 is the logarithm of 1000 to the base 10 or, symbolically,

$$3 = \log_{10}1000.$$

Since exponentiation is not commutative and a^b and b^a are generally not equal it follows that these two inverse operations, root extraction and anti-exponentiation (sometimes also called "logarithm taking"), are *essentially different* operations.

In view of our extended definition of exponentiation (to include fractional exponents) the root extraction problem of finding a number a such that

$$a^b = c$$

is identical with the exponentiation problem of finding a number a such that

$$c^{1/b} = a.$$

Thus if we make no distinction between exponentiation with integral exponents and exponentiation with fractional exponents, then root extraction becomes essentially the same operation as exponentiation. (Thus when we cite the failure of the rational numbers to include such numbers as $2^{1/2}$, we can consider this as a lack of closure for the inverse operation of root extraction or for the operation of exponentiation itself.)

1.4.4.4. *Evidence that the rational number system is not closed under exponentiation*

Our troubles with the rational number system all started with a Greek fellow named Pythagoras who, when he should have been digging for clams like everyone else, spent his time drawing pictures in the sand instead. Pythagoras and some of his fellow Greek sand-artists found more or less regular diversion in drawing diagrams in which certain sets of points were connected by **segments** (or pieces) of straight lines. In order to make this seem sufficiently important to be a respectable substitute for clam digging they called their activity **geometry,** meaning earth measurement.†

Pythagoras was particularly interested in a configuration involving three points (not all lying on any one straight line) and the three line segments joining them. Pythagoras called such a configuration a **triangle** (or "three-angle"), the word **angle** meaning the space at one of the three points enclosed between the two line segments meeting there. Of all triangles, the ones found most interesting were those which contained one **right angle,** a right angle being one which contains (between the two line segments forming it) exactly one fourth of all the space surrounding the common point of these two line segments. The three line segments which form a triangle are called the **sides** of the triangle. In a right triangle the longest side always lies opposite the right angle and is called the **hypotenuse** of the triangle. The following figure shows the nature of a right triangle and its various parts.

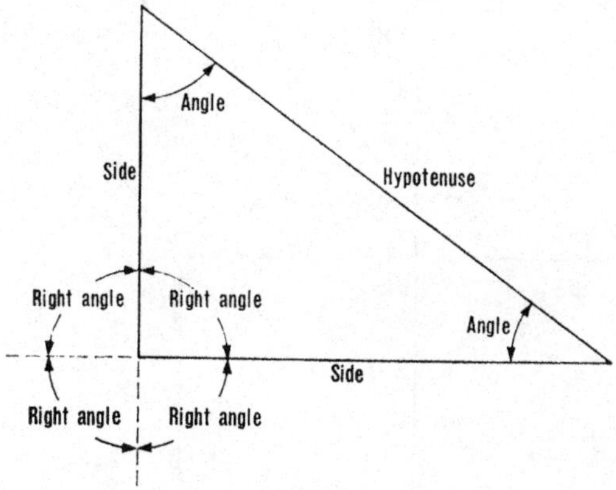

One of the things which Pythagoras found most fascinating about right triangles was the relationship between the three

† Mathematicians are like that. They think big.

squares which could be constructed by using the sides of a right triangle as sides (of the squares). Suppose we use the letter c to represent the length of the hypotenuse of a right triangle and the letters a and b to represent the lengths of the other two sides. Then, since everyone knew by this time that the area of a square was the product obtained by multiplying the length of a side of the square by itself, the areas of these three squares would be $c \cdot c$, $a \cdot a$ and $b \cdot b$, or c^2, a^2 and b^2 respectively.

The relationship discovered by Pythagoras (and subsequently *proved* by him on the basis of good mathematical logic) was that, for *any* and *every* right triangle, the area c^2 of the square constructed on the hypotenuse was *exactly* equal to the sum of the areas a^2 and b^2 of the squares constructed on the other two sides, or that

$$a^2 + b^2 = c^2.$$

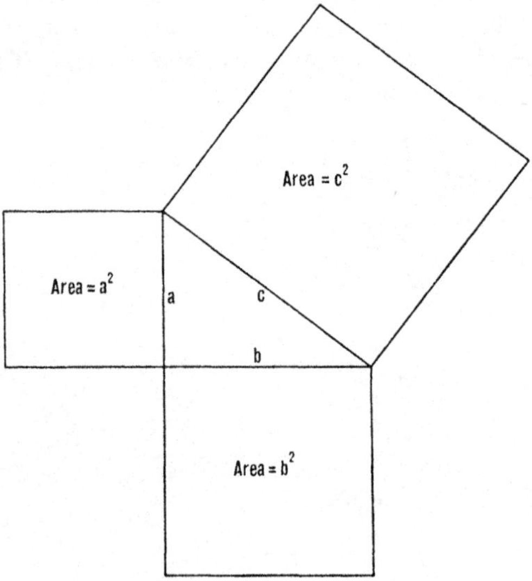

The discovery of this surprising and interesting fact not only delighted Pythagoras and his friends and contemporaries, but has also proved to be a most gratifyingly useful bit of information to the entire mathematically interested segment of humanity ever since its discovery and, to this day, not a word has ever been said against Pythagoras for frittering away his time in this manner instead of helping his wife to provide clams for the table. However, something that did get him into very nearly serious trouble with District 50 of the United Mathematical Workers was this: if Pythagoras' law be applied to a right triangle in which each of the sides a and b is equal to one, then the hypotenuse c is such that

$$c^2 = 1^2 + 1^2 = 1 + 1 = 2;$$

that is, c is a number whose square (or second power) is two, and hence c is the principal square root of two, or

$$c = 2^{1/2}.$$

While this, in itself, does not appear objectionable, attempts to express $2^{1/2}$ as a rational number did produce results which caused Pythagoras' mathematical brotherhood to view the situation with alarm. Being accustomed, as they were, to regarding the rational number system as perfectly adapted to describing all measurable physical or geometrical quantities, they were greatly perplexed to learn that $2^{1/2}$ did not appear to be a rational number. As a matter of fact, it was found that **$2^{1/2}$ was indeed not a rational number.** This fact can be proved by the following argument.

Let us formally assume, as did Pythagoras and his contemporaries, that $2^{1/2}$ is a rational number, say p/q, where p and q are integers. Since any fraction *can* be reduced to its lowest terms, let us assume that this has already been done in the case of p/q so that *p and q are integers which have no common integral divisor.*

If such an assumption be tenable, then it would follow that

$$(2^{1/2})^2 = \left(\frac{p}{q}\right)^2$$

or

$$2 = \frac{p^2}{q^2}$$

whence

$$2 \cdot q^2 = p^2.$$

This equality is a formal statement that $2 \cdot q^2$ and p^2 are the same number. Then since 2 is a factor, or exact divisor, of $2 \cdot q^2$, it follows that 2 is a factor of p^2.

Now either 2 is a factor of p and hence p is of the form $2 \cdot k$ for some integer k, or 2 is not a factor of p and then p is necessarily of the form $2 \cdot k + 1$ for some integer k. (There is no third alternative.) But if p were of the form $2 \cdot k + 1$, then p^2 would be of the form

$$\begin{aligned} p^2 &= (2k+1)^2 = (2k+1) \cdot (2k+1) = (2k) \cdot (2k+1) \\ &\quad + 1 \cdot (2k+1) \\ &= (2k) \cdot (2k) + (2k) \cdot 1 + 1 \cdot (2k) + 1 \cdot 1 = 4k^2 + 2k \\ &\quad + 2k + 1 \\ &= 4k^2 + 4k + 1 = 2 \cdot (2k^2 + 2k) + 1. \end{aligned}$$

Since $2k^2 + 2k$ is an integer (due to the closure of the system of integers under both multiplication and addition), it follows that $2 \cdot (2k^2 + 2k) + 1$, and hence p^2, is *not* exactly divisible by 2. Since this is a direct contradiction of the previously entertained proposition that p^2 is divisible by 2, and since this contradiction arose from the tentative assumption that p was of the form $2 \cdot k + 1$, it follows that this assumption was erroneous and hence that p is of the form $2 \cdot k$, and hence divisible by 2. Then

$$p^2 = (2 \cdot k)^2 = 4 \cdot k^2$$

and the equality

$$2 \cdot q^2 = p^2$$

becomes

$$2 \cdot q^2 = 4 \cdot k^2$$

or

$$q^2 = 2 \cdot k^2.$$

Then since 2 is a factor of $2 \cdot k^2$, it is a factor of q^2 and, by an argument exactly similar to the preceding one, 2 is a factor of q so that q is of the form $2 \cdot h$, for some integer h.

We have now shown (on the basis of the assumption that $2^{1/2}$ can be expressed in the rational form p/q where p and q are integers with no common integral factor) that p and q contain the common integral factor 2, which is, of course, a direct contradiction. This contradiction proves the erroneousness of our assumption that $2^{1/2}$ could be expressed in the form of a rational number. We can only conclude, therefore, that $2^{1/2}$ is not a rational number.

Pythagoras and his colleagues, being good organization men, tried to solve the problem of their embarrassment caused by this defect in what they had previously regarded as a perfect number system by hushing the matter up as far as the general public was concerned. However, "the truth will inevitably out," and it was only a matter of time before the whole disgraceful matter was known to everyone. As a matter of fact it was found that $3^{1/2}$, $5^{1/2}$, $6^{1/2}$, $7^{1/2}$, and many other square roots were not rational numbers. In general, it was found that if any integer was not the square of some integer then it was not the square of any rational number, and hence its square root was not a rational number. Similar things could be said about certain cube roots, fourth roots, fifth roots, etc. Another disconcerting fact was that the number called π (pi)† was not a rational number. As a result of all these crushing blows the mathematical world found itself in a state of severe shock for some time.

1.4.4.5. *The real numbers*

After the initial shock, shame and chagrin wore off and people reconciled themselves to the inadequacy of the rational number system it was immediately obvious what should be done about it. The system should be expanded again to take in all those numerical concepts like $2^{1/2}$, $3^{1/2}$, $2^{1/3}$, π, etc., which had physical or geometrical significance but which did not fit into the rational

† The Greek letter π is used to denote the number (discovered by ancient Greek mathematicians) which describes the length of the circumference of a circle whose diameter is one unit in length. This number is also equal to the *ratio* of the length of the circumference of any circle to the length of its diameter.

number system. Since many, at least, of these numbers could be represented along with the rational numbers by points on a number scale (that is, a number-labeled straight line) a seemingly natural solution to the problem was to invent a new number system in which there would be one number corresponding to each point on such a line. Since such number concepts as $2^{1/2}$, $3^{1/2}$, π, etc., corresponded to certain dimensions in the real physical world, the new system of numbers was called the **real number system.** The straight line number scale which inspired the invention of this system is frequently referred to as "the real number line" or, more simply, "the real line".

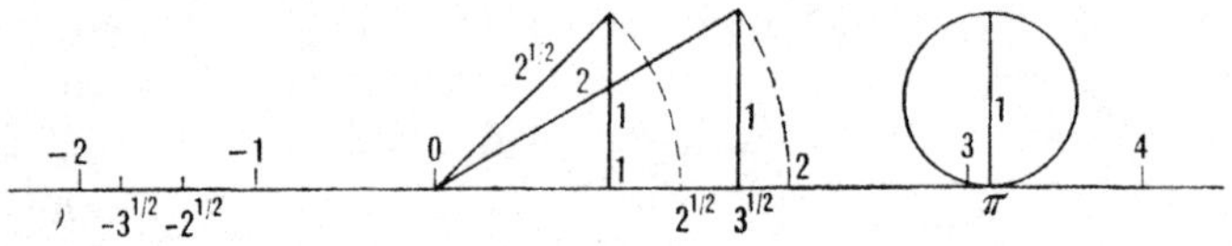

All real numbers corresponding to points on the same side of the real line as those points corresponding to positive rational numbers are defined as positive; those corresponding to points on the other side (where the negative rational numbers are located) are defined as negative. The discussion used to demonstrate the addition and subtraction of positive and negative integers (by eastward and westward pacings) is equally plausible with positive and negative real numbers. Previous definitions and consequent rules for multiplication, subtraction, division and exponentiation may, with appropriate modification in interpretation, be found to be consistent and suitable for use with real numbers. While a thorogoing proof of this last statement would be so lengthy and tedious that we shall not attempt it, it is a fact to which we can certify that the real number system is closed under addition, subtraction, multiplication, division and certain aspects of exponentiation and its inverse operations. In it the operations of addition and multiplication are commutative and associative and there exists an identity element for each of these operations. Multiplication is distributive over addition and subtraction, division is one-sidedly distributive over division and subtraction, and exponentiation is one-sidedly distributive over multiplication and division. All real numbers which are not rational are called **irrational numbers.**

Now that $2^{1/2}$ and other such non-rational expressions have been accorded formal number status the next thing to do is to find a suitable standard way of finding and expressing the value of such numbers so that comparisons can be made involving these and other real numbers. To do this we make use of the following adaptations of the rational numbers.

1.4.4.6. *Decimal representation*

We are accustomed to the fact that the principle of positional notation allows us to express any integer as a sum of integral multiples of integral powers of ten. Thus the numeral 27,643 is interpreted to mean

$$2 \cdot 10^4 + 7 \cdot 10^3 + 6 \cdot 10^2 + 4 \cdot 10^1 + 3 \cdot 10^0$$

or

$$2 \cdot 10^4 + 7 \cdot 10^3 + 6 \cdot 10^2 + 4 \cdot 10 + 3.$$

A very similar method can be developed for representing any fraction which is less than one (we call such fractions **proper** fractions) if the denominator of the fraction is a product of powers of two and/or five. First of all, any such fraction can be rewritten in an equivalent (or *equal-valued*) form in which the denominator is an integral power of ten. The method of accomplishing this is illustrated in the following example:

$$\frac{183}{250} = \frac{183}{2 \cdot 5^3} = \frac{2^2 \cdot 183}{2^2 \cdot (2 \cdot 5^3)} = \frac{4 \cdot 183}{2^3 \cdot 5^3} = \frac{732}{10^3}.$$

Such a fraction may then be written as a sum of integral multiples of negative integral powers of ten, thus:

$$\frac{732}{10^3} = \frac{7 \cdot 10^2 + 3 \cdot 10 + 2}{10^3} = \frac{7 \cdot 10^2}{10^3} + \frac{3 \cdot 10}{10^3} + \frac{2}{10^3}$$
$$= 7 \cdot 10^{-1} + 3 \cdot 10^{-2} + 2 \cdot 10^{-3}.$$

When we add an integer such as 27,643 and a proper fraction such as 183/250 the resulting sum 27,643 + 183/250 is called a **mixed number** and is commonly written (without the addition symbol +) as

$$27{,}643\,\frac{183}{250}.$$

Using the above results, this mixed number can now be written as

$$2 \cdot 10^4 + 7 \cdot 10^3 + 6 \cdot 10^2 + 4 \cdot 10^1 + 3 \cdot 10^0 + 7 \cdot 10^{-1} + 3 \cdot 10^{-2} + 2 \cdot 10^{-3}.$$

Since the use of positional notation (with base ten) works so well for the integral part of this mixed number it is natural to want to extend its use to the fractional part also. We do this by writing

$$27{,}643\,\frac{183}{250} = 27{,}643.732,$$

where the fractional part is represented by .732 and the dot . (called a **decimal point** when used in this manner) is used as a boundary marker to separate the integral part of the mixed number from the fractional part.† A proper fraction written in a form such as .732 is called a **decimal fraction.** The mixed number written in the form 27,643.732 exemplifies the **decimal form** for writing a general rational number.

For a fraction such as 19/37 (which *cannot* be written in an equivalent form with a denominator which is a power of ten) the matter of obtaining a decimal representation is still possible but somewhat more complicated.

It may be found by trial that 19/37 is greater than 5/10 and less than 6/10, and by further trial that it is greater than $5/10 + 1/10^2$ and less than $5/10 + 2/10^2$, and so on. However, it is not necessary to use trial-and-error methods, as information of this sort

† It is interesting to note that while the decimal point is printed at the same level as a period in the United States, in England it is printed at a slightly higher level.

can be obtained in a more direct and businesslike manner by the following method:

$$\frac{19}{37} = \frac{1}{10}\cdot\frac{190}{37} = \frac{1}{10}(5+\frac{5}{37}) = \frac{5}{10}+\frac{1}{10}\cdot\frac{5}{37} = \frac{5}{10}+\frac{1}{10^2}\cdot\frac{50}{37}$$

$$= \frac{5}{10}+\frac{1}{10^2}(1+\frac{13}{37}) = \frac{5}{10}+\frac{1}{10^2}+\frac{1}{10^2}\cdot\frac{13}{37}$$

$$= \frac{5}{10}+\frac{1}{10^2}+\frac{1}{10^2}\cdot\frac{130}{37} = \frac{5}{10}+\frac{1}{10^2}+\frac{1}{10^3}(3+\frac{19}{37})$$

$$= \frac{5}{10}+\frac{1}{10^2}+\frac{3}{10^3}+\frac{1}{10^3}\cdot\frac{19}{37}.$$

This shows that 19/37 is greater than $5/10 + 1/10^2 + 3/10^3$ and less than $5/10 + 1/10^2 + 4/10^3$. Also, since we now have at least *some* information on the decimal expression of 19/37, we can use it to obtain further information in the following manner:

$$\frac{19}{37} = \frac{5}{10}+\frac{1}{10^2}+\frac{3}{10^3}+\frac{1}{10^3}\cdot\frac{19}{37}$$

$$= \frac{5}{10}+\frac{1}{10^2}+\frac{3}{10^3}+\frac{1}{10^3}(\frac{5}{10}+\frac{1}{10^2}+\frac{3}{10^3}+\frac{1}{10^3}\cdot\frac{19}{37})$$

$$= \frac{5}{10}+\frac{1}{10^2}+\frac{3}{10^3}+\frac{5}{10^4}+\frac{1}{10^5}+\frac{3}{10^6}+\frac{1}{10^6}\cdot\frac{19}{37}$$

$$= \frac{5}{10}+\frac{1}{10^3}+\frac{3}{10^3}+\frac{5}{10^4}+\frac{1}{10^5}+\frac{3}{10^6}$$

$$+\frac{1}{10^6}(\frac{5}{10}+\frac{1}{10^2}+\frac{3}{10^3}+\frac{1}{10^3}\cdot\frac{19}{37})$$

$$= \frac{5}{10}+\frac{1}{10^2}+\frac{3}{10^3}+\frac{5}{10^4}+\frac{1}{10^5}+\frac{3}{10^6}+\frac{5}{10^7}+\frac{1}{10^8}+\frac{3}{10^9}$$

$$+\frac{1}{10^9}\cdot\frac{19}{37}, \text{ etc.}$$

It is thus seen that the decimal form for the fraction 19/37 involves an unending sequence of repetitions of the digits 5, 1 and 3 like this:

$$\frac{19}{37} = .513513513\ldots,$$

where the three-dot symbol (called an **ellipsis**) is to be interpreted as meaning that the established pattern of repetition continues without termination. Such a decimal expression is called a **repeating non-terminating decimal.**† (For contrast, consider the **terminating decimal** .732 which represents the fraction 183/250.) While the non-terminating decimal shown above represents the rational number 19/37, we shall find that **it is possible to represent any real number whatever, rational or not, by some decimal,** either terminating, non-terminating and repeating, or non-terminating and non-repeating.

For example, since

$$1^2 = 1, \text{ which is less than } 2,$$
$$(2^{1/2})^2 = 2,$$

and

$$2^2 = 4, \text{ which is greater than } 2,$$

we reason that $2^{1/2}$ is greater than 1 and less than 2. Also, since

$$1.4^2 = 1.96, \text{ which is less than } 2,$$

and

$$1.5^2 = 2.25, \text{ which is greater than } 2,$$

we reason that $2^{1/2}$ is greater than 1.4 and less than 1.5. In the same way we find that

† In a repeating decimal the repeating sequence of digits is called the **repetend.** As a convenient shorthand method of representing a repeating decimal, the repetend may be written *just once with a bar drawn above it* immediately following the non-repeating part of the number. For example $23\frac{19}{37} = 23.513513513\ldots = 23.\overline{513}$; $7\frac{2}{3} = 7.66666\ldots = 7.\overline{6}$; etc. (Note, however, that the same meaning would be represented by $23.513513\overline{513}$, $23.513\overline{513}$, $23.513513513\overline{513}$, $23.\overline{513}$, $23.5\overline{135}$, $23.51\overline{351}$, etc.)

1.41	$<$	$2^{1/2}$	$<$	1.42
1.414	$<$	$2^{1/2}$	$<$	1.415
1.4142	$<$	$2^{1/2}$	$<$	1.4143
1.41421	$<$	$2^{1/2}$	$<$	1.41422
1.414213	$<$	$2^{1/2}$	$<$	1.414214
1.4142135	$<$	$2^{1/2}$	$<$	1.4142136
1.41421356	$<$	$2^{1/2}$	$<$	1.41421357
1.414213562	$<$	$2^{1/2}$	$<$	1.414213563, etc.,

where, it will be recalled, the statement $1.41 < 2^{1'2} < 2.42$, for instance, means "$2^{1/2}$ is *greater than* 1.41 and *less than* 1.42". From the above comparison of $2^{1/2}$ with certain terminating decimals representing rational numbers we can make the following corresponding comparison of differences.

$2^{1/2} - 1$	$< 2.0 - 1.$	$= 1.$
$2^{1/2} - 1.4$	$< 1.5 - 1.4$	$= .1$
$2^{1/2} - 1.41$	$< 1.42 - 1.41$	$= .01$
$2^{1/2} - 1.414$	$< 1.415 - 1.414$	$= .001$
$2^{1/2} - 1.4142$	$< 1.4143 - 1.4142$	$= .0001$
$2^{1/2} - 1.41421$	$< 1.41422 - 1.41421$	$= .00001$
$2^{1/2} - 1.414213$	$< 1.414214 - 1.414213$	$= .000001$
$2^{1/2} - 1.4142135$	$< 1.4142136 - 1.4142135$	$= .0000001$
$2^{1/2} - 1.41421356$	$< 1.41421357 - 1.41421356$	$= .00000001$
$2^{1/2} - 1.414213562$	$< 1.414312563 - 1.414213562$	$= .000000001$,

etc.

Thus we see that there is a sequence of rational numbers

1.
1.4
1.41
1.414
1.4142
1.41421
1.414213
1.4142135
1.41421356
1.414213562, etc.,

whose succeeding terms differ from $2^{1/2}$ by progressively smaller amounts in such a way that "by going sufficiently far out" in the sequence we could find a term which would differ from $2^{1/2}$ by less than any small amount that we might care to specify. This circumstance is described in formal terminology by saying that "the sequence **converges** to the number $2^{1/2}$ as a **limit**".

In exactly the same way, it is seen that the sequence

.5
.51
.513
.5135
.51351
.513513
.5135135
.51351351
.513513513, etc.,

converges to the number 19/37 as a limit.

Thus we have, as examples, two *infinite* (or *unending*) sequences of rational decimal numbers, one of which converges to the irrational number $2^{1/2}$ as a limit and the other of which converges to the rational number 19/37 as a limit. We notice that the decimal expression

$$1.414213562 \ldots$$

for the irrational number $2^{1/2}$ is a *non-terminating* and *non-repeating* decimal. (The three-dot symbol in this case means merely an unending sequence of digits of some sort, since no pattern of repetition has been established.) On the other hand, we notice that the decimal expression for the rational number 19/37 is a *non-terminating repeating* decimal. In both cases the behavior is characteristic since it can be proved that **the decimal expression for any irrational number must be non-terminating and non-repeating whereas the decimal expression for any rational number must be either terminating or non-terminating and repeating.**

In both of the preceding examples we have considered a convergent sequence of rational numbers. There are, of course, also sequences of irrational numbers. Some of these converge to rational limits and some converge to irrational limits. As examples, the sequence of irrational numbers

$$(2/1)^{1/2}, (3/2)^{1/2}, (4/3)^{1/2}, \text{etc.},$$

converges to the rational limit 1, and the sequence of irrational numbers

$$(1/2)\pi, (2/3)\pi, (3/4)\pi, \text{etc.}$$

converges to the irrational number π.

Two important conclusions emerge from the foregoing discussion. One is that **the real number system is closed under the operation of convergence to a limit.** (It should be noticed that this is *not* a binary operation but a new kind of operation involving an infinite sequence of numbers.) The other is that **every real number can be expressed as a decimal, either terminating or non-terminating.** Since a number which can be expressed by means of a positional system with one base can be expressed also by means of a positional system with any other base (any integer greater than one will serve as a base) it follows that any real number can also be expressed in a binary (base two) system of notation, an octal (base eight) system, a duodecimal (base twelve) system, or any other.

1.4.4.7. *Scientific notation*

While any real number can be represented in a decimal form, some decimal representations—even terminating ones—are somewhat awkward to write. This is obviously the case for very large numbers (such as 29,700,000,000.) and for very small numbers (such as .000000000000372).

Scientists, engineers and others who deal with numbers of this sort frequently have adopted a shorthand method for expressing them which reduces the writing of all such numbers to a compact and relatively simple standardized form. This style of writing a number is called **scientific notation** and consists of expressing it as a product of an integral power of ten multiplied by a multiplier which is less than ten but not less than one. We illustrate the method by the following applications to the two numbers mentioned in the preceding paragraph.

$$29{,}700{,}000{,}000. = 2.97 \cdot 10{,}000{,}000{,}000. = 2.97 \cdot 10^{10}$$
$$.000000000000372 = 3.72 \cdot .0000000000001$$
$$= 3.72 \cdot 1/10{,}000{,}000{,}000{,}000.$$
$$= 3.72 \cdot 1/10^{13} = 3.72 \cdot 10^{-13}.$$

EXERCISES

1. Evaluate each of the following powers as a number expressed in the simplest possible form.
(a) 17^0 (b) $(5/8)^{-1}$ (c) 10^{-5} (d) $16^{1/4}$ (e) $4^{-1/2}$ (f) $.001^{2/3}$ (g) $.0000000001^{1/5}$ (h) $64^{-5/6}$ (i) 3^{-4} (j) $216^{1/3}$.

2. Write the following numbers in scientific notation.
(a) 393,000. (b) .0156. (c) .0000109 (d) 24,300,000,000,000,000.

3. Write the following numbers in ordinary decimal form.
(a) $6.82 \cdot 10^{-7}$ (b) $2.10 \cdot 10^{18}$ (c) $5.05 \cdot 10^{6}$ (d) $9.01 \cdot 10^{-14}$.

4. Prove that the number $3^{1/2}$ (the "square root of three") is not rational.

5. At the bottom of page 121 there are given the first ten numbers of an infinite sequence of rational numbers which converges to $2^{1/2}$. Find the *first six* numbers of a similar sequence which converges to $3^{1/2}$.

6. At the top of page 122 there are given the first nine numbers of an infinite sequence of rational numbers which converges to 19/37. Find the *first eight* numbers of a similar sequence which converges to 716/3333.

7. Determine the rational number represented by the repeating decimal $.\overline{2434}$.

8. What procedure can you devise for shortening and streamlining the method of pages 119 and 120 for finding the decimal representation for a rational number given in fractional form?

9. Is the operation of root extraction commutative? Is it associative? Is it distributive over addition? Is it distributive over multiplication? Does it have an identity element? Does it have an inverse operation?

10. Apply (and answer) the questions asked in Exercise 9 for the operation of anti-exponentiation.

11. List ten real numbers which are not rational.

12. Give an example of the application of one of our familiar binary operations to two real numbers where the result is not a real number.

13. Show, by means of either a diagram or an outline, the inclusion relations between all of the different kinds of numbers that we have discussed so far.

14. Representing the set of all natural numbers by N, the set of all counting numbers by C, the set of all integers by I, the set of all rational numbers by

R and the set of all real numbers by L, write the proper formal set-inclusion relations for each pair of these sets. Are your results consistent with those you obtained in Exercise 13?

15. Are there more rational or irrational numbers? Explain your reasoning.

16. Express the following fractions in the binary analog of the decimal notation: (a) 1/8; (b) 1/64; (c) 1/3; (d) 13/16; (e) 2/5.

17. Express the following fractions in the octal analog of the decimal notation: (a) 3/8; (b) 37/128; (c) 1/3; (d) 2/5; (e) 11/13.

18. Prove that multiplication is not distributive over exponentiation.

19. Prove that exponentiation is not distributive over addition.

20. Prove that addition is not distributive over exponentiation.

21. It has been shown that $(a \cdot b)^c = a^c \cdot b^c$. *Derive* a similar equivalent form for $(a/b)^c$.

22. Show that if an operation Δ is not left-distributive for addition then neither is it left-distributive for subtraction.

23. Is the set of irrational numbers closed under (a) addition? (b) multiplication? (c) subtraction? (d) division?

1.4.5. Root Extraction and the Determination of Logarithms; the Complex Numbers

1.4.5.1. *Inadequacy of the real number system*

Since even a "terminating" decimal such as .732 can be written in a repeating non-terminating form like

$$.732000000\ \ldots\ = .732\overline{0},$$

we can summarize the findings of the preceding section by saying that **every real number, rational or irrational, can be expressed in the form of a non-terminating decimal.**

We have noted that the non-terminating decimal

$$1.414213562\ \ldots$$

(which expresses the value of $2^{1/2}$) is the limit to which the infinite sequence 1, 1.4, 1.41, 1.414, 1.4142, 1.41421, 1.414213, 1.4142135, 1.41421356, 1.414213562, ... converges. The principal virtue of the real number system is that whenever an infinite sequence of real numbers converges to a limit, the limit is itself a real number. That is, the real number system is closed with respect to the convergence of infinite sequences. It is clearly illustrated by the case of $2^{1/2}$ that the rational number system does not have this property, since the numbers 1, 1.4, 1.41, 1.414, etc., are all rational numbers but the limit, $2^{1/2}$, is not a rational number.

Our extended definition of powers makes clear the meaning of such powers as $2^{1/2}$, $2^{1/3}$, $5^{1/2}$, $10^{5/8}$, etc., and we have seen that by introducing the real numbers and the use of non-terminating decimals we can evaluate such powers in a manner suitable for practical reckoning. We are now in a position to extend our definition of powers still further and define not only powers with rational exponents but also powers with irrational exponents such as, for instance, $10^{(2^{1/2})}$. To do this, we first determine an infinite sequence of rational numbers such as

1, 1.4, 1.41, 1.414, 1.4142, 1.41421, 1.414213, 1.4142135, 1,41421356, etc.,

or, equivalently,

$$1, \frac{14}{10}, \frac{141}{100}, \frac{1,414}{1,000}, \frac{14,142}{10,000}, \frac{141,421}{100,000}, \frac{1,414,213}{1,000,000}, \frac{14,142,135}{10,000\ 000}, \frac{141,421,356}{100,000,000}, \text{etc.,}$$

which converges to the proposed exponent $2^{1/2}$. We then form the corresponding sequence of powers with rational exponents:

$$10^{1}$$
$$10^{14/10}$$
$$10^{141/100}$$
$$10^{1,414/1,000}$$
$$10^{14,142/10,000}$$
$$10^{141,421/100,000}$$
$$10^{1,414,213/1,000,000}$$
$$10^{14,142,135/10,000,000}$$
$$10^{141,421,356/100,000,000}, \text{ etc.}$$

The limit to which this sequence converges is *defined* to be the value of

$$10^{(2^{1/2})}.$$

(Note that each of the above powers with rational exponent is **a number that can be determined as a decimal.** For instance, $10^{141,421,356/100,000,000}$ is merely the 100,000,000th root of the 141,421,356th power of ten.)

We have now shown how to interpret and evaluate the power a^b for any real number b and any positive number a. We also know how to interpret and evaluate this power if a is zero and b is any positive real number; and we can also interpret and evaluate it if a is any negative real number and b is any integer or any rational number whose denominator is an odd integer. For example:

$$
\begin{aligned}
&(2/3)^{2/3} = [(2/3)^2]^{1/3} = [4/9]^{1/3} \\
&0^{17} = 0 \\
&0^{1/5} = 0 \\
&0^{3/11} = 0 \\
&0^{\pi} = 0 \\
&(-4)^4 = (-4)\cdot(-4)\cdot(-4)\cdot(-4) = 256 \\
&(-4)^3 = (-4)\cdot(-4)\cdot(-4) = -64 \\
&(-8)^{1/3} = -2 \\
&(-8)^{2/3} = (-2)^2 = 4 \\
&(-8)^0 = 1, \text{ etc.}
\end{aligned}
$$

One case still not covered by our extended definition, however, is that of a power with a negative base and a rational exponent whose denominator is an even integer. For example, the power

$$(-1)^{1/2}$$

cannot be interpreted as any number in the real number system. This makes it clear that the real number system is still not closed under general exponentiation (or its inverse, root extraction). It should be realized that this is not a trivial matter since the successful practice of modern engineering technology frequently demands that we be able to give a practical interpretation of the above "square root of the negative of the number one" as a number. This necessity having been established, our duty is clear; we must again enlarge the number system to include $(-1)^{1/2}$

and all other powers provided that we do not admit any combination of base and exponent which would imply division by zero (such as 0^{-2}) or which would lead to a meaningless form (such as 0^0).

1.4.5.2. *The enlargement of the number system to provide closure for general exponentiation*

In order to get ideas on just how to go about creating a still further expanded number system to include such "numbers" as $(-1)^{1/2}$ we might take advantage of previous experience (after all, we've been thru this sort of thing often enough before) and see what happens if we treat $(-1)^{1/2}$ as a number. By "treating it as a number", we mean that we shall assume that it obeys all applicable laws with respect to all the natural operations and their inverses that are valid in the real number system and its predecessors. Along with this, of course, we must remember that $(-1)^{1/2}$ has the distinction that its square (or second power) is the number -1.

One of the important results of the peculiar nature of $(-1)^{1/2}$ (which we shall represent by the symbol "i" from now on) is that it has only four distinct integral powers. This is seen as follows:

$$\begin{aligned}
i^1 &= i \\
i^2 &= [(-1)^{1/2}]^2 = (-1)^{(1/2)2} = (-1)^1 = -1 \\
i^3 &= i^2 \cdot i = (-1) \cdot i = -i \\
i^4 &= i^3 \cdot i = (-i) \cdot i = -i^2 = -(-1) = 1 \\
i^5 &= i^4 \cdot i = 1 \cdot i = i \\
i^6 &= i^4 \cdot i^2 = 1 \cdot (-1) = -1 \\
i^7 &= i^4 \cdot i^3 = 1 \cdot (-i) = -i \\
i^8 &= (i^4)^2 = 1^2 = 1 \\
i^9 &= i^8 \cdot i = 1 \cdot i = i, \text{ etc.}
\end{aligned}$$

As a consequence of this limitation on the number of distinct powers of i, every expression involving a sum of multiples of powers of i can be reduced to a very simple form similar to the one in the following example:

$$\begin{aligned} 7 + 5 \cdot i - 2 \cdot i^2 + 14 \cdot i^3 + 10 \cdot i^4 + i^5 - 19 \cdot i^6 \\ = 7 + 5 \cdot i - 2 \cdot (-1) + 14 \cdot (-i) + 10 \cdot 1 + i - 19 \cdot (-1) \\ = 7 + 5 \cdot i + 2 - 14 \cdot i + 10 + i + 19 \\ = 7 + 2 + 10 + 19 + 5 \cdot i - 14 \cdot i + i \\ = (7 + 2 + 10 + 19) + (5 - 14 + 1) \cdot i \\ = 38 + (-8) \cdot i \text{ (or, more simply, } 38 - 8i). \end{aligned}$$

Thus, except for simplifications such as the ones just performed, it is never necessary to work with any power of i beyond the first, and all sums of multiples of powers of i can be simplified to a form such as

$$a + bi$$

for some real numbers a and b. All expressions of this form are called **complex numbers.**

If, in the complex number $a + bi$, the real number b is zero, then the complex number reduces to the real number a (since we assume $0 \cdot i = 0$). If b is different from zero, then $a + bi$ is called an **imaginary number.** If b is different from zero and $a = 0$, then $a + bi$ reduces to the form bi which is called a **pure imaginary number.**†

The term "imaginary" as used in this connection is an outgrowth of the difficulty which the first workers with the complex numbers had in trying to "picture" a square root of the negative of the number one, and in accepting it as a "number". The word "imaginary" should not be taken too literally or seriously nowadays since the use of the complex numbers has come to play quite as important a part in modern science as the so-called "real" numbers.

Now let's see what sort of characteristics the complex numbers have *if* (as we shall assume) *all the usual natural laws of commutativity, associativity and distributivity apply.*

Suppose that a, b, c and d are any real numbers so that $a + bi$ and $c + di$ are any two complex numbers. Then

† This is an American usage. In England the term "imaginary" conveys the same meaning as the one which we have given here for "pure imaginary".

$$\begin{aligned}(a+bi)+(c+di) &= a+(bi+c)+di\\ &= a+(c+bi)+di\\ &= (a+c)+(bi+di)\\ &= (a+c)+(b+d)i.\end{aligned}$$

Since the real numbers are closed under addition, $a + c$ and $b + d$ are real numbers and hence the sum $(a + c) + (b + d)i$ is a complex number. Thus (if our assumption of the other laws be accepted) **the complex numbers are closed under addition.** Similarly,

$$(a+bi)-(c+di) = (a-c)+(b-d)i$$

and **the complex numbers are closed under subtraction.** Also

$$\begin{aligned}(a+bi)(c+di) &= ac+a(di)+(bi)c+(bi)(di)\\ &= ac+(ad)i+(bc)i+(bd)i^2\\ &= ac+(ad+bc)i+(bd)(-1)\\ &= ac+(ad+bc)i-bd\\ &= (ac-bd)+(ad+bc)i.\end{aligned}$$

By the closure of the real numbers under multiplication, addition and subtraction $ac - bd$ and $ad + bc$ are real numbers and hence the product $(ac - bd) + (ad + bc)i$ is a complex number. Thus **the complex numbers are closed under multiplication.** To show that they are also closed under division, we must show that the indicated quotient $(a + bi)/(c + di)$ is also a complex number. To do this we use the trick of multiplying both numerator and denominator by the complex number $c - di$ (which is called the **conjugate** of $c + di$).

$$\begin{aligned}\frac{a+bi}{c+di} &= \frac{(a+bi)(c-di)}{(c+di)(c-di)} = \frac{ac-bdi^2+(bc-ad)i}{c^2-d^2i^2}\\ &= \frac{ac-bd(-1)+(bc-ad)i}{c^2-d^2(-1)}\\ &= \frac{ac+bd+(bc-ad)i}{c^2+d^2}\\ &= \frac{ac+bd}{c^2+d^2}+\frac{bc-ad}{c^2+d^2}i.\end{aligned}$$

Since the real numbers are closed under multiplication, addition, subtraction and division, it follows that $(ac + bd)/(c^2 + d^2)$ and $(bc - ad)/(c^2 + d^2)$ are both real numbers and hence the quotient

$$\frac{ac + bd}{c^2 + d^2} + \frac{bc - ad}{c^2 + d^2}i$$

is a complex number. Hence **the complex numbers are closed under division.**

It can also be shown (altho not so simply—so we shall not do it) that **the complex numbers are closed under exponentiation and its inverses** so that **if any two of the symbols *a*, *b* and *c* represent complex numbers then the third will also represent some complex number for which the relation**

$$a^b = c$$

is true.

It can also be shown (altho not simply) that the concept of convergence can be defined for infinite sequences of complex numbers and that **the complex numbers are closed under convergence**; that is, whenever an infinite sequence of complex numbers converges to a limit, the limit is also a complex number.

Even if we had to stop with this listing of the virtues of the complex number system, it would be a marvelously complete and useful number system. As a matter of fact, however, it has still other surprising and gratifying properties. In order to appreciate one of the most important of these it is helpful to understand the concept of a polynomial. A "**polynomial in the number x**" **is a sum of multiples of powers of *x* involving only positive integers (and zero) as exponents.** For example, the expression

$$7 + 5i - 2i^2 + 14i^3 + 10i^4 + i^5 - 19i^6$$

which we discussed a few pages back is a "polynomial in *i*", In particular, we call this a polynomial "of the **sixth degree in *i***" because six is the largest exponent used with the base *i*. Similarly.

$$2x - 3$$

is a polynomial of the **first degree in *x***,

$$ay^2 + by + c$$

is a polynomial of the **second degree in y**,

$$3z^3 - 19z^2 + 14z + 23$$

is a polynomial of the **third degree in z**,

$$z^4 - 22z + 1$$

is a polynomial of the **fourth degree in z**,

$$62k^{16} - 34k^{11} + 29k^8 + k^4 - 33k$$

is a polynomial of the **sixteenth degree in k**, etc. In general, **for any non-negative integer n, a polynomial of the nth degree in the number x is any expression of the form**

$$a_0x^n + a_1x^{n-1} + a_2x^{n-2} + \dots + a_{n-2}x^2 + a_{n-1}x + a_n$$

where the symbols $a_0, a_1, a_2, \dots, a_{n-2}, a_{n-1}, a_n$ (read as "a-sub-zero", "a-sub-one", "a-sub-two", etc.) **are to be understood as representing any particular set of $n + 1$ numbers (provided a_0 is not zero).**

A statement of the form

$$a_0x^n + a_1x^{n-1} + a_2x^{n-2} + \dots + a_{n-2}x^2 + a_{n-1}x + a_n = 0,$$

which states that the value of an nth degree polynomial is the number zero, is called an "algebraic equation of degree n". It should be noted that a statement of the form

$$a_0x^n + a_1x^{n-1} + \dots + a_{n-1}x + a_n = b_0x^n + b_1x^{n-1} + \dots + b_{n-1}x + b_n$$

which states that two nth degree polynomials are equal is also an algebraic equation of degree n since it can be written in the form

$$(a_0 - b_0)x^n + (a_1 - b_1)x^{n-1} + \dots + (a_{n-1} - b_{n-1})x + (a_n - b_n) = 0.$$

The formation of an nth degree polynomial may be considered as an $(n + 2)$-nary operation; that is, the combination of the $n + 2$ numbers $x, a_0, a_1, a_2, \dots, a_n$ to form a new number. Each of the number systems is closed for this operation. The inverse operation of finding one of the numbers $a_0, a_1, a_2, \ldots, a_n$ when all the others, the number x, and the value of the polynomial are known is trivially simple since it involves only addition, subtraction, multiplication and division. Thus, if

$$a^7 x_0 + a_1 x^6 + a_2 x^5 + a_3 x^4 + a_4 x^3 + a_5 x^2 + a_6 x + a_7 = b,$$

where $a_0, a_1, a_2, a_3, a_5, a_6, a_7$, x and b are known, then

$$a_4 = \frac{b - (a_0 x^7 + a_1 x^6 + a_2 x^3 + a_3 x^4 + a_5 x^2 + a_6 x + a_7)}{x^3}.$$

A much more interesting inverse operation (because it is generally much more difficult) is the operation of finding a number x such that

$$a_0 x^n + a_1 x^{n-1} + a_2 x^{n-2} + \ldots + a_{n-2} x^2 + a_{n-1} x + a_n = b$$

when $a_0, a_1, a_2, \ldots, a_{n-2}, a_{n-1}, a_n$ and b are known. This operation is called **solving** the equation and each value of x found for which the equation is a **true** statement is called **a solution** of the equation. Note that since this equation can be written in the form

$$a_0 x^n + a_1 x^{n-1} + a_2 x^{n-2} + \ldots + a_{n-2} x^2 + a_{n-1} x + (a_n - b) = 0$$

the process of solving it is essentially of the same nature as that for solving the simpler appearing equation

$$a_0 x^n + a_1 x^{n-1} + a_2 x^{n-2} + \ldots + a_{n-2} x^2 + a_{n-1} x + a_n = 0;$$

hence we may standardize the discussion of all nth degree algebraic equations by referring to the latter simple form.

If the number n is one, the equation becomes

$$a_0 x + a_1 = 0$$

and the process of solving it is quite simple, leading to the solution

$$x = -\frac{a_1}{a_0}.$$

If n is two, the equation becomes

$$a_0x^2 + a_1x + a_2 = 0$$

and the solutions can be shown without much difficulty to be

$$x = \frac{-a_1 + (a_1{}^2 - 4a_0a_2)^{1/2}}{2a_0} \text{ and } x = \frac{-a_1 - (a_1{}^2 - 4a_0a_2)^{1/2}}{2a_0}.$$

If n is three, the equation becomes

$$a_0x^3 + a_1x^2 + a_2x + a_3 = 0$$

and it can be shown that there are three solutions, each given by a formula involving a_0, a_1, a_2 and a_3 and in which both square roots and cube roots are involved.

If n is four, the equation becomes

$$a_0x^4 + a_1x^3 + a_2x^2 + a_3x + a_4 = 0$$

and it can be shown that it has four solutions, each of which may be expressed in terms of a_0, a_1, a_2, a_3 and a_4 by means of addition, subtraction, multiplication, division and the extraction of square and cube roots.

It has been proved that **the general algebraic equation of degree n, in which the coefficients (or multipliers (a_0, a_1, a_2, ..., a_n may be any complex numbers (provided $a_0 \neq 0$), has exactly n solutions (altho these may not all be distinct), and each solution is a complex number.** Hence **the complex numbers are closed under the operation of solving algebraic equations.** This is a property not possessed by any of the number systems which we have previously considered, and is one which makes the complex number system not merely most useful but indispensable for purposes of general scientific analysis and technological computation.

Unfortunately, for our convenience, it has also been proved that for n greater than four, the solutions of algebraic equations generally cannot be expressed precisely by formulas involving only a finite (or limited) number of additions, subtractions, multiplications, divisions and root extractions.

1.4.6. Graphical Comparison of the Number Systems

The natural numbers can be represented by a set of uniformly paced points on a half-line, thus,

and the counting numbers by the same diagram with the starting point labeled to represent the number zero.

The integers can be represented by a similar set of uniformly spaced points on a full line.

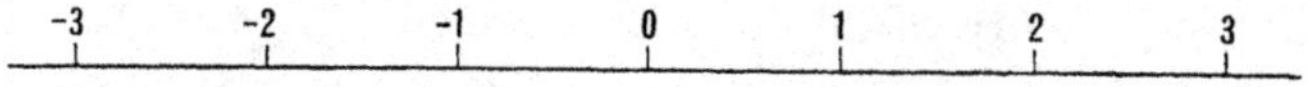

To represent the rational numbers we start with the scheme of integer-points on a full line, add to this all points which are an integral number of half-units from the zero point (a **unit** being the distance between the points labeled zero and one), then add all points which are an integral number of one-third units from the zero point, then all one-fourth unit points, then all one-fifth unit points, and so on.

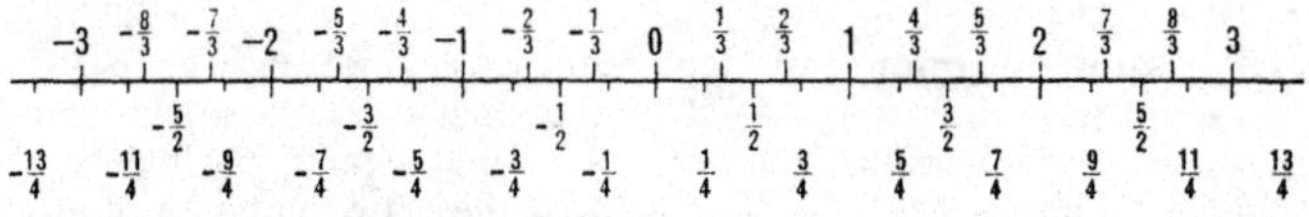

It might seem that the points representing all the rational numbers would fill the entire line but, surprisingly, this is not the case because we recall that there are points corresponding to numbers such as $2^{\frac{1}{2}}$, $3^{\frac{1}{2}}$, etc., which are not rational.

The real numbers are represented by *all* the points on the line.

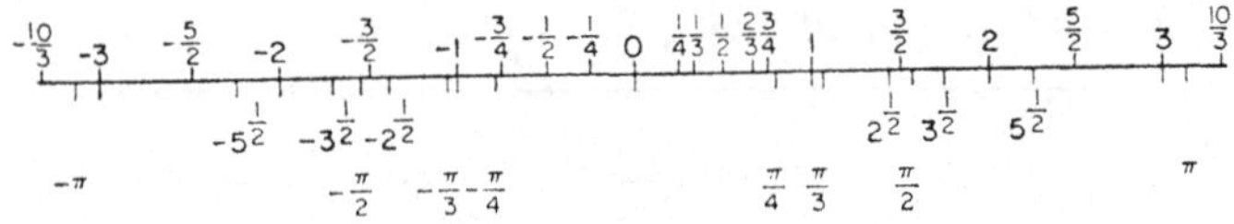

To represent the complex numbers (each of which involves two real numbers) we use two real number scales placed at right angles to each other (one horizontal, one vertical) with their zero points coinciding. A complex number $a + bi$ is then represented by a point which lies vertically opposite the real number a on the horizontal line and horizontally opposite the real number b on the vertical line. Two real numbers such as a and b, when used in this manner to locate a point, are called "the **coordinates**" of that point. The totality of all complex numbers corresponds to the totality of all points in the plane of the two number scales. A plane used in this manner to represent the complex number system is called "the complex plane". The horizontal scale is called "the axis of real numbers" (or simply "the real axis") and the vertical scale is called "the axis of pure imaginary numbers" (or "the imaginary axis"). The common zero point on the two number scales is called the "**origin of coordinates**" or, sometimes, simply "the **origin**".

Exercises

1. On a sheet of graph paper (preferably "engineering paper") draw a suitable set of axes and, using one-fifth of an inch as a unit, show the point A which represents the complex number $3 + 4i$ and the point B which represents the complex number $12 + 5i$. The sum of these two complex numbers is

$$(3 + 4i) + (12 + 5i) = (3 + 12) + (4 + 5)i = 15 + 9i\,.$$

Show the point S which represents this sum. Draw the line segment OA, OB, AS and BS where O is the origin (or common zero point on both scales). What kind of figure is $OASB$? State a brief conjecture (limited to one sentence) on a graphical method for adding complex numbers.

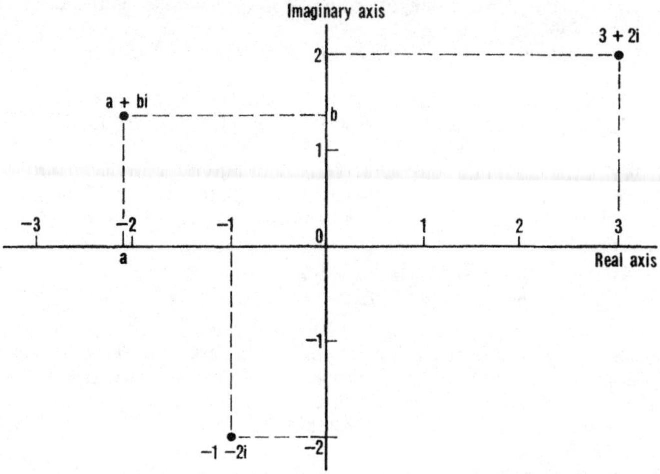

2. Represent the complex numbers $3 + 4i$ and $12 + 5i$ as in the preceding problem. (Use one-tenth inch as one unit for this problem.) Their product is

$$\begin{aligned}(3 + 4i)(12 + 5i) &= 3 \cdot 12 + 3 \cdot (5i) + (4i) \cdot 12 + (4i) \cdot (5i)\\ &= 36 + 15i + 48i + 20i^2\\ &= 36 + 63i + 20(-1)\\ &= 36 - 20 + 63i\\ &= 16 + 63i.\end{aligned}$$

Show the point P which represents this product. Measure the distances OA, OB and OP. State a brief conjecture (limited to one sentence) concerning the relation of OP to OA and OB. Measure the angles ROA, ROB and ROP (where R is a point on the right side of the real axis). State a brief conjecture (limited to one sentence) concerning the relation of angle ROP to angles ROA and ROB.

3. If $z_1 = 5 + 2i$ and $z_2 = 29 - 29i$, find (in simplest form): (a) $z_1 + z_2$; (b) $z_1 - z_2$; (c) $z_2 - z_1$; (d) $z_1 \cdot z_2$; (e) z_1/z_2; (f) z_2/z_1.

4. Let the symbol ε represent the complex number $(3^{1/2} + i)/2$. Compute ε^2, ε^3, ε^4, ε^5, ε^6, ε^7, ε^8, ε^9, ε^{10}, ε^{11} and ε^{12}, taking advantage of the earlier results to simplify the work of obtaining the later ones. Draw suitable axes on a sheet of graph paper and, using two inches as one unit, show the number ε and each of its computed powers.

5. List the conjugate for, and use it to find the reciprocal of, each of the following complex numbers: (a) $3 - 4i$; (b) $-6 - 2i$; (c) $5i$; (d) 19. (e) What is the conjugate of zero?

6. (a) Plot on graph paper the points representing the complex number $3 + 4i$, its negative, its conjugate, and the negative of its conjugate. (b) Prove

that the negative of the conjugate of any complex number is the same as the conjugate of its negative.

7. Find a complex number (different from one) which is equal to its own fourth power.

8. The fifth degree equation $x^5 - 32 = 0$ has the solution $x = 32^{1/5}$, or $x = 2$. Does this contradict the last paragraph in section 1.4.5.2 (on page 134)? Explain.

9. We have assumed the laws of commutativity and associativity for addition and multiplication and the law of distributivity of multiplication over addition for the complex numbers and have shown that the following relations were consequences:

$$(a + bi) + (c + di) = (a + c) + (b + d)i$$
$$(a + bi) \cdot (c + di) = (ac - bd) + (ad + bc)i.$$

Some people prefer to state these latter two relations as the *definitions* of addition and multiplication for complex numbers and then, using these definitions and the known commutative, associative and distributive laws for the real numbers, *prove as consequences* that the corresponding laws are also valid for the complex numbers. Do this for the case of commutativity of addition.

10. Do as in Exercise 9 for associativity of addition.

11. Do as in Exercise 9 for commutativity of multiplication.

12. Do as in Exercise 9 for associativity of multiplication.

13. Do as in Exercise 9 for distributivity of multiplication over addition.

14. (a) What complex number is the identity element for addition in the system of complex numbers? (b) What complex number is the identity element for multiplication in the system of complex numbers? Justify your answers.

15. Can you find two complex numbers u and v such that $u = v^2$ and $v = u^2$?

16. Is the set of imaginary numbers closed under (a) addition? (b) multiplication? (c) subtraction? (d) division?

17. Is the set of pure imaginary numbers closed under (a) addition? (b) multiplication? (c) subtraction? (d) division?

CHAPTER 1.5

CLASSIFICATION OF THE NUMBER SYSTEMS AS ABSTRACT SYSTEMS

THE foregoing discussion has been concerned with the properties of the number systems and operations of arithmetic from an intuitive viewpoint.

While there cannot be much doubt that this viewpoint was the dominant one in the historical development of mankind's awareness of these concepts, present-day mathematicians tend to consider it more fruitful to study these concepts not with regard to the intuitive aspects of their individual properties but rather on the basis of the overall patterns formed collectively by all such properties. These patterns may be used to define certain concepts called **abstract systems** of which the various number systems may be recognized as constituting special cases. We now endeavor to explain this way of looking at the number systems.

1.5.1. Inverse Elements

We have discussed the nature of inverse operations. In dealing with abstract systems, however, the more precisely defined concept of **inverse element** is more extensively employed.

In section 1.3.5 we have defined an identity element for a binary operation Δ in a set S of elements to be an element e in S such that for each element s of S it is true that

$$e \Delta s = s \text{ and } s \Delta e = s.$$

In the set of counting numbers, the set of integers, the set of rational numbers, the set of real numbers and the set of complex numbers the element zero is an identity element for the operation of addition. In the set of natural numbers, the counting numbers, the integers, the rationals, the reals and the complexes, the element one is an identity element for the operation of multiplication.

If a set S contains an identity element e for a binary operation Δ and if for some particular element—say s_1—in S, there exists an element—say t_1—such that

$$s_1 \Delta t_1 = e \text{ and } t_1 \Delta s_1 = e,$$

then t_1 is called the **inverse element of s_1 for Δ.** (From the nature of the definition it follows also that s_1 is the inverse element of t_1 for Δ.)

It may be noted that, in the set of integers (or in any other number system which contains the integers), **the inverse element of any element for addition is its negative,** and that, in the set of rational numbers (or any other number system which contains this set), **the inverse element of any non-zero element for multiplication is its reciprocal.**

Since there is no identity element for exponentiation (in any number system) the question of the existence of inverse elements of given elements for this operation is not applicable.

The relation between inverse operations and inverse elements (where both exist) is illustrated by the examples

$$a + (-b) = a - b$$

and

$$a \cdot (1/b) = a/b.$$

From these it may be induced that, in general, **the application of a given operation to the inverse of an element yields the same result as the application of the inverse operation to the element itself.** Symbolically, if Δ is a binary operation and $\triangledown$ its inverse and if a, b and b' are elements with b and b' each the inverse of the other, then

$$a \Delta b' = a \triangledown b.$$

1.5.2. Abstract Systems

For our present purpose we shall define **an abstract system** as **a set of elements for which one or more binary operations is defined.** We distinguish one abstract system from another by means of the further special properties which they may have.

The simplest example of an abstract system among our number systems is the system of natural numbers with the binary operation of addition. The characteristic properties of this system are that it is **closed,** the operation (of addition) is **associative** for all elements in the set, and the operation is **commutative** for all elements in the set.

If we enlarge our system to the counting numbers by adjoining the single new number zero it still enjoys the three properties just mentioned and has also the additional distinction of possessing an **identity element** since zero is the identity element for the operation of addition.

When our number system is further enlarged to include all the integers, then it possesses not only the four properties listed above but also the interesting new property that for each element in the new system there is also present within the system the **additive inverse** (or negative) element corresponding to it.

Altho we have been concentrating our attention on the operation of addition, it is also true that the operation of multiplication is defined on the set of natural numbers, the set of counting numbers and the set of integers. If we ignore the operation of addition and concentrate on the systems formed by each of these sets with the operation of multiplication, then in each case we have the properties of **closure, associativity, commutativity,** and the existence of an **identity element** (the number one).

If we consider the system formed by the set of integers and **both** the operations of addition and multiplication, then, of course, we have all the properties listed above for these individual operations and also the property of **distributivity of multiplication over addition.**

When we adjoin the fractions to the integers to obtain the set of rational numbers and consider the system formed by this set with the two operations of addition and multiplication, then we have all the properties already listed and one new one, namely the existence of a **multiplicative inverse** for each non-zero element in the set. It should be noticed that the systems formed by either the real numbers or the complex numbers with the two operations of addition and multiplication enjoy all these same properties.

The observant and number-conscious reader may have noted that we have now mentioned a total of **eleven** properties. It is these eleven properties which are used in defining the abstract systems which we shall presently define and of which the various number systems may be recognized as special cases.

To define—by the names which have been accepted as customary for them—the abstract systems which are of importance to us in connection with the number systems, let us consider a non-empty set A of elements for which there is defined at least one binary operation ①, and possibly a second binary operation ② also, and which possesses one or more of the following properties:

1. A is closed for ①.
2. ① is associative for all elements in A.
3. There exists an identity element for ① in A.
4. For each element in A the inverse of this element for ① is also in A.
5. ① is commutative for all elements in A.
6. A is closed for ②.
7. ② is associative for all elements in A.
8. There exists an identity element for ② in A.
9. For each element in A (other than the identity element for ①) the inverse of this element for ② is also in A.
10. ② is commutative for all elements in A.
11. ② is distributive over ① for all elements in A.

Then the abstract system formed by the set A with ① (or ① and ②) is called:

a **groupoid** if it possesses property 1;
a **semi-group** if it possesses properties 1 and 2;

a **monoid** if it possesses properties 1, 2, and 3;
a **GROUP** if it possess properties 1, 2, 3 and 4;
a **module** (or **commutative group,** or **Abelian group**) if it possesses properties 1, 2, 3, 4 and 5;
a **ring** if it possesses properties 1, 2, 3, 4, 5, 6, 7 and 11;
a **ring with unity element** if it possesses properties 1, 2, 3, 4, 5, 6, 7, 8 and 11;
a **commutative ring** if it possesses properties 1, 2, 3, 4, 5, 6, 7, 10 and 11;
a **commutative ring with unity element** if it possesses properties 1, 2, 3, 4, 5, 6, 7, 8, 10 and 11;
an **integral domain** if it is a commutative ring with unity element and for any two elements a and b in A, a ② $b = 0$ only if either $a = 0$ or $b = 0$ (where 0 is the identity element for ①;
a **skew-field** if it possesses properties 1, 2, 3, 4, 5, 6, 7, 8, 9, and 11;
a **FIELD** if it possesses properties 1, 2, 3, 4, 5, 6, 7, 8, 9, 10 and 11.

One might be forgiven, possibly, if after a lapse of time the exact nature of a groupoid, a semi-group, a monoid or a module were to elude one's memory. To fail to retain a clear and vivid mental image of either a group or a field, however, would be a much more serious matter since **the group is perhaps the most basically important of all abstract mathematical systems and the field is a concept of pre-eminent importance in considerations involving the more extensive number systems.** It is interesting to note that applications of the group concept have been found in quantum physics and other branches of modern science.

Since an abundance of examples of the various abstract systems may be cited without difficulty among the common number systems, we shall reserve these for the exercises and not dwell on them here. Instead, to exemplify the idea of a group in general, let us consider some groups whose elements are not numbers.

1.5.3. Examples of Groups

One of the most instructive group categories is that of the **permutation groups.** A **permutation** (as we shall use the term here) shall be understood to mean **the act of permuting (or changing) an ordered arrangement of a fixed number of things into another (or, possibly, back into the same) ordered arrangement of the same things.** For any fixed number of things, the set of all permutations of them forms a group. The binary operation for this group is called **composition** and consists of combining two permutations (to obtain a third one) by observing the final result produced after applying first one permutation and then applying the second to the result of the first.

We shall illustrate the idea of a permutation group by considering the group of permutations on three things. The three "things" could be any three things whatever—a clam, a brontosaurus and a pterodactyl—or an orange, a lemon and a lime—or the numbers 1, 2 and 3. We shall use the latter three solely for the reason that they are easiest to write.

To represent that permutation, or change, in the arrangement of these three numbers (whatever this original arrangement may be) in which 1 is replaced by 2, 2 is replaced by 3, and 3 is replaced by 1 we shall write the symbol (123). The permutation in which 1 is replaced by 3, 3 by 2, and 2 by 1 would be represented by (132). When 1 is replaced by 2, 2 by 1, and 3 is replaced by itself (that is, 3 is left unchanged), we omit any mention of the item which replaces itself and write just (12). When each of the three items replaces itself we write just (1)—since we must write something to represent the situation. Thus the six possible permutations are:

(123), meaning 1 is replaced by 2, 2 by 3, and 3 by 1;
(132), meaning 1 is replaced by 3, 3 by 2, and 2 by 1;
(12), meaning 1 is replaced by 2, 2 by 1, and 3 by 3;
(13), meaning 1 is replaced by 3, 3 by 1, and 2 by 2;
(23), meaning 2 is replaced by 3, 3 by 2, and 1 by 1;
(1), meaning 1 is replaced by 1, 2 by 2, and 3 by 3.

A good way to appreciate the nature of these permutations is to observe their effect on various ordered arrangements of digits such as, for instance, the social security number 512-42-2613, the year number 1963, the street address 2123, the price $32.95, the license number 36K-143, etc. When the permutation (123) is applied to the number 512-42-2613 the latter is changed to 523-43-3621 and we indicate this symbolically by writing

(123) [512-42-2613] = 523-43-3621.

Similarly,

(132) [512-42-2613] = 531-41-1632,
(132) [523-43-3621] = 512-42-2613,
(1) [512-42-2613] = 512-42-2613,
(12) [512-42-2613] = 521-41-1623,
(13) [512-42-2613] = 532-42-2631,
(23) [512-42-2613] = 513-43-3612,
(12) [523-43-3621] = 513-43-3612,
(123) [1963] = 2961,
(132) [2123] = 1312,
(23) [2123] = 3132,
(123) [$32.95] = $13.95,
(123) [36K-143] = 16K-241, etc.

The result of **composing** two of these permutations (that is, of following one by the other) is called their **composition product.** The composition product of (123) followed by (132) is written as (132)(123) and is equal to (1) since in (123) 1 is replaced by 2 and in (132) 2 is replaced by 1 so that the net result of applying (123) and then (132) is to replace 1 by 1; similarly, in (123) 2 is replaced by 3 and in (132) 3 is replaced by 2 so that ultimately 2 is replaced by itself; likewise 3 is replaced by 1 and then 1 is replaced by 3 so that 3 replaces itself in the final result also. Thus we write

$$(132)(123) = (1)$$

because

$$\begin{aligned}(132)(123)[512\text{-}42\text{-}2613] &= (132)[(123)[512\text{-}42\text{-}2613]]\\ &= (132)\ [523\text{-}43\text{-}3621]\\ &= [512\text{-}42\text{-}2613]\\ &= (1)\ [512\text{-}42\text{-}2613]\end{aligned}$$

and because similarly equal results occur when the composition product (132)(123) and the permutation (1) are both applied to any numeral whatever. For the same reason we write

$$\begin{aligned}(123)(123) &= (132),\\ (12)(123) &= (23), \text{ etc.}\end{aligned}$$

The results of all such compositions may be compiled in the form of a table (which might be called the **operation table** for the group or, in this particular example of a group, the **composition table**) as follows, the first permutation being shown at the top of the table and the one which follows it (but is *written before* it) being shown in the column at the left of the table.

	(1)	(12)	(13)	(23)	(123)	(132)
(1)	(1)	(12)	(13)	(23)	(123)	(132)
(12)	(12)	(1)	(132)	(123)	(23)	(13)
(13)	(13)	(123)	(1)	(132)	(12)	(23)
(23)	(23)	(132)	(123)	(1)	(13)	(12)
(123)	(123)	(13)	(23)	(12)	(132)	(1)
(132)	(132)	(23)	(12)	(13)	(1)	(123)

The information given in this table is most useful and convenient for verifying the group properties (properties 1, 2, 3 and 4) for our set of six permutations. Property 1 (closure) is verified by the fact that only the six permutations themselves appear in the body of the table. Property 3 (existence of an identity element) is verified by the fact that the entire row in the body of the table opposite (1) is identical with the row at the top, and the entire column in the body of the table under (1) is identical with the column at the left; this shows that (1) is the identity element. Property 4 (existence of inverses) is shown by the fact that for each permutation in the column at the left there is some (unique) corresponding permutation at the top of the table such that the composition product of the two, shown in the body of the table, is the identity element (1). (Another way of saying the last sentence is that the identity element occurs once, and only once, in each row and in each column of the table. It may be noticed that a similar statement is true for each of the permutations in the group; this is characteristic of all group operation tables.) Verification of property 2 (associativity) by means of the table would involve detailed comparison of all 216 triple composition products of the form $p_1(p_2p_3)$ with the corresponding ones of the form $(p_1p_2)p_3$—where the symbols p_1, p_2 and p_3 represent some three of the six permutations. There is no denying that this would be tedious.

A better way to prove (quickly and easily) that composition of permutations is associative is to make the following use of the two principles: (1) two permutations, say p and p', are **equal** if, and only if, they have identical effects when applied to **any** numeral N, and (2) the composition product of any two permutations is that permutation which has the same effect on any numeral N as that effect produced by applying the two permutations one after the other to the same numeral N. That is, in symbols,

(1) $p = p'$ if, and only if, $p[N] = p'[N]$, (for **any** numeral N),
(2) $(pp')[N] = p[p'[N]]$.

Using these principles, we have, for any three permutations p_1, p_2, p_3 and **any** numeral N:

$$(p_1(p_2p_3))[N] = p_1[(p_2p_3)[N]) = p_1[p_2[p_3[N]]$$

and also

$$((p_1p_2)p_3)[N] = (p_1p_2)[p_3[N]] = p_1[p_2[p_3[N]].$$

Thus, since $(p_1(p_2p_3))[N] = ((p_1p_2)p_3)[N]$ for **any** numeral N it follows that the permutations $p_1(p_2p_3)$ and $(p_1p_2)p_3$ are equal and hence composition of permutations is associative.

Similar groups (and composition tables) exist, of course, for the set of all permutations of four things, the set of all permutations of five things, and in general, the set of all permutations of n things, where n may be any positive integer.

Another interesting category of groups is that of the rotation groups of the regular polyhedra† (the tetrahedron, cube, octahedron, dodecahedron and icosahedron). The elements of these groups are those **rotations** which leave a regular polyhedron in a position congruent to its original one (that is, in a position in which it occupies precisely the same region of space as before altho individual vertices, edges and faces need not all be in their respective original positions). By recognizing that each such rotation involves a definite *permutation of vertices* of the regular polyhedron in question, the description and discussion of the rotations and their group properties can be handled quite simply in the manner of the preceding example by merely numbering the vertices.

For instance, in the case of the regular tetrahedron, we number the vertices with the numbers 1, 2, 3 and 4 and then consider the set of permutations of these numbers which represent possible rotations of a tetrahedron.

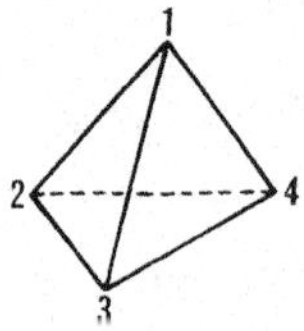

The permutation (124) is a possible one, for example, since if the tetrahedron be rotated clockwise thru an angle of 120° about an axis passing thru vertex 3 and the center of the opposite triangular face, vertex 1 will be replaced by vertex 2, 2 will be replaced by 4, 4 will be replaced by 1, and vertex 3 will remain fixed. Likewise, the permutation (14)(23) represents a possible rotation—namely a

† If you don't know what these are, ask someone to explain them—even your instructor. It's a legitimate question.

rotation of the tetrahedron thru 180° about an axis passing thru the midpoint of the edge joining vertices 1 and 4 and the midpoint of the edge joining 2 and 3. The permutation (14), however, does not represent a possible rotation since it is impossible to rotate the tetrahedron in such a manner as to interchange vertices 1 and 4 while at the same time leaving each of vertices 2 and 3 fixed. The following list of permutations represent all the possible congruent rotations and may be shown to form a group.

(1)	(124)	(143)	(12)(34)
(123)	(142)	(234)	(13)(24)
(132)	(134)	(243)	(14)(23)

A third example of a group (called a **substitution group**) consists of the six elements

$$r,\ \frac{1}{r},\ 1 - r,\ 1 - \frac{1}{r},\ \frac{1}{1-r},\ \frac{r}{r-1}$$

and the binary operation of **substitution** of any one of these expressions for the symbol r in any one of the expressions. Thus, representing this operation by the symbol Ⓢ, the "substitution product" of $1/r$ and $r/(r - 1)$ would be

$$\frac{1}{r} \text{Ⓢ} \frac{r}{r-1} = \frac{(1/r)}{(1/r) - 1} = \frac{1}{1-r}$$

and the commuted substitution product of $r/(r - 1)$ and $1/r$ would be

$$\frac{r}{r-1} \text{Ⓢ} \frac{1}{r} = \frac{1}{r/(r-1)} = \frac{r-1}{r} = 1 - \frac{1}{r}.$$

It may be noted that the substitution operation Ⓢ is not commutative.

Exercises

1. In answering the following questions specify both the set and the appropriate operation (or operations) which satisfy the required conditions. (For example: The natural numbers form a groupoid with the operation of addition; the integers form a ring with the operations of addition and multiplication; etc.) From among the number systems we have studied, list all the: (a) groupoids; (b) semi-groups; (c) monoids; (d) groups; (e) modules; (f) rings; (g) commutative rings; (h) rings with unity elements; (i) commutative rings with unity elements; (j) integral domains; (k) skew-fields; (l) fields.

2. Cite an example of a group which is not commutative.

3. Cite an example of a ring without a unity element.

4. Define: (a) a semi-group in terms of a groupoid;
(b) a monoid in terms of a semi-group;
(c) a group in terms of a monoid;
(d) a module in terms of a group;
(e) a ring in terms of a module;
(f) an integral domain in terms of a ring;
(g) a field in terms of an integral domain.

5. Define a field as concisely as possible in terms of the group concept.

6. List all the possible permutations of the four numbers 1, 2, 3 and 4.

7. (a) Is the group of permutations of three things commutative? (b) Is the group of permutations of four things commutative? (c) Is the group of permutations of two things commutative?

8. Show that the two possible "triple-products" of the three permutations (1234), (1432) and (413), taken in that order, are equal.

9. How many pairs of "triple-products" would have to be compared (as in Exercise 8) to verify the associativity property for the group of all permutations of four things?

10. Draw a diagram of a cube, number the vertices, and then list all permutations of the vertices that represent possible congruent rotations of the cube.

11. Draw a diagram of a regular octahedron, number the vertices, and then list all permutations of the vertices that represent possible congruent rotations of the octahedron.

12. Construct the operation table for the rotation group of the regular tetrahedron.

13. Construct the operation table for the substitution group with elements r, $1/r$, $1-r$, $1-(1/r)$, $1/(1-r)$, $r/(r-1)$.

14. Construct the multiplication table for the set of complex numbers $\{1, i, -1, -i\}$. What system does this set form with the operation of multiplication?

15. Construct the multiplication table for the set of complex numbers $\{1, -1/2 + 3^{1/2}i/2, -1/2 - 3^{1/2}i/2\}$. What system does this set form with the operation of multiplication?

16. Give an example of a monoid whose elements are not numbers.

17. What system is formed with addition and multiplication by the set of all even integers?

1.5.4. Examples of Fields

It can be verified that certain of the standard number systems we have discussed are fields. Since this verification is a routine matter, however, we shall reserve our attention here for examples of fields which are somewhat more unusual.

EXAMPLE 1. FIELD OF SURDS

A number of the form $a + k^{1/2}b$, where a and b are rational numbers and k is an integer which is not a perfect square, is called a **surd.** Let's consider the set of all surds with $k = 2$, that is, the set of all numbers of the form $a + 2^{1/2}b$ where a and b may be any rational numbers. Call this set S.

Let $a + 2^{1/2}b$ and $c + 2^{1/2}d$ be any two surds in S. Then

$$(a + 2^{1/2}b) + (c + 2^{1/2}d) = (a + c) + 2^{1/2}(b + d)$$

and this sum is also in S since, by the closure of the rational numbers for addition, $a + c$ and $b + d$ are rational numbers. Thus S is closed under addition. Similarly, the product

$$(a + 2^{1/2}b) \cdot (c + 2^{1/2}d) = (ac + 2bd) + 2^{1/2}(ad + bc)$$

is also in S because, due to the closure of the rational numbers under both addition and multiplication, $ac + 2bd$ and $ad + bc$ are rational numbers. Thus S is closed under multiplication. It can be shown to follow as a consequence of the associative, commutative and distributive properties of the rational numbers that S has the corresponding properties. The identity elements for addition and multiplication in the rational number system serve in the same respective capacities for S. Finally, for each element in S, its additive inverse can be shown to be in S, and for each non-zero element in S, its multiplicative inverse can also be shown to be in S. Thus S is a field.

Similar fields exist for surds of the form $a + k^{1/2}b$, where k may be any non-square integer.

EXAMPLE 2. FIELD OF CONGRUENCE CLASSES

In that branch of mathematics known as number theory, it is customary to say that two integers are "**congruent** modulo m" (where m is a specified positive integer) if the difference of the two integers is an integral multiple of m. For example, 3 and 18 are congruent modulo 5 since

$$18 - 3 = 15 = 3 \cdot 5.$$

The integer m is called a **modulus** and the word "modulo" means "to the modulus" or "with respect to the modulus". The congruence relation is somewhat similar to the equality relation but is not the same as equality and should not be confused with it. The statement "3 is congruent to 18, modulo 5" is written symbolically as

$$3 \equiv 18, (5).$$

The set of all integers that are mutually congruent with respect to a particular modulus is called a **congruence class** (or a **residue class**) for that modulus. For example, the set of all integers of the form $3 + 5k$, where k is an integer, is a congruence class for the modulus 5. This set may be written in various ways, including the following:

$\{\ldots, -12, -7, -2, 3, 8, 13, 18, 23, \ldots\}$;
$\{3 + 5k \mid k \text{ is an integer}\}$;
$[3]$.

The last notation being the simplest (because it was designed specifically for the purpose of being simple) is obviously the most convenient to use. This class is called "the congruence class represented by 3". The other congruence classes for the modulus 5 are [0], [1], [2] and [4]. In general, for any modulus m, there are m distinct congruence classes, namely [0], [1], [2], [3], ..., $[m - 1]$.

In view of the definition of congruence it follows that any integral multiple of the modulus is congruent to zero. As a consequence (for the modulus m),

$$[m] = [0]$$
$$[m + 1] = [1]$$
$$[m + 2] = [2], \text{ etc.}$$

Thus all *distinct* congruence classes for the modulus m can be represented by integers in the set $\{0, 1, 2, 3, \ldots, m - 1\}$.

Operations called "addition" and "multiplication" may be defined for congruence classes by the following relations:

$$[p] + [q] = [p + q]$$
$$[p] \cdot [q] = [p \cdot q].$$

This means that, in adding the congruence class represented by p and the congruence class represented by q, the sum is defined to be the congruence class represented by $p + q$. Similarly, in multiplying the congruence class represented by q by the congruence class represented by p, the product is defined to be the congruence class represented by $p \cdot q$. Thus, for example, the congruence classes modulo 5 may be added and multiplied as follows:

$$[1] + [2] = [1 + 2] = [3]$$
$$[2] + [3] = [2 + 3] = [5] = [0]$$
$$[3] + [4] = [3 + 4] = [7] = [2], \text{ etc.};$$

$$[1] \cdot [2] = [1 \cdot 2] = [2]$$
$$[2] \cdot [3] = [2 \cdot 3] = [6] = [1]$$
$$[3] \cdot [4] = [3 \cdot 4] = [12] = [2], \text{ etc.}$$

Proceeding in this manner, we establish the following addition and multiplication tables for the congruence classes modulo 5.

+	[0]	[1]	[2]	[3]	[4]
[0]	[0]	[1]	[2]	[3]	[4]
[1]	[1]	[2]	[3]	[4]	[0]
[2]	[2]	[3]	[4]	[0]	[1]
[3]	[3]	[4]	[0]	[1]	[2]
[4]	[4]	[0]	[1]	[2]	[3]

·	[0]	[1]	[2]	[3]	[4]
[0]	[0]	[0]	[0]	[0]	[0]
[1]	[0]	[1]	[2]	[3]	[4]
[2]	[0]	[2]	[4]	[1]	[3]
[3]	[0]	[3]	[1]	[4]	[2]
[4]	[0]	[4]	[3]	[2]	[1]

From these two tables it may be verified that the set {[0], [1], [2], [3], [4]} consisting of the five congruence classes modulo 5 satisfies each of the eleven axioms for a field; therefore this set of congruence classes forms a field with the operations of "addition" and "multiplication" as defined for them. It is interesting to note that, whereas there are infinitely many elements in any number fields of the type previously discussed or considered, there are *only five* elements in this field.

Sometimes, among those who are familiar with this "modular arithmetic", the brackets which we have used as congruence class symbols are omitted and the five elements of this field are written simply as 0, 1, 2, 3 and 4. Accordingly, we write

$$3 + 2 = 0$$
$$3 + 3 = 1$$
$$3 \cdot 3 = 4, \text{ etc.}$$

At first this sort of arithmetic, in which a non-zero number (5 in our example) is treated like zero, seems very strange and unreasonable. One's feeling of distrust of modular arithmetic and reluctance to have anything to do with it is usually dissipated rather quickly, however, when one is reminded that this is exactly the sort of arithmetic which we have used all our lives in keeping track of time by clocks and watches—with the number twelve as the modulus. Right? Recall also that when adding angles whose sums exceed 360°, we also frequently use modular arithmetic with 360° as the modulus. So modular arithmetic is not as strange as it first seems—in fact, no stranger than "telling time" or adding angles.

EXERCISES

1. Show that the set S of all surds of the form $a + 2^{1/2}b$ (where a and b are rational) is closed under (a) subtraction; (b) division.

2. Find (a) the additive inverse, and (b) the multiplicative inverse, of $5 + 2^{1/2} \cdot 2$. Find (c) the additive inverse, and (d) the multiplicative inverse, of $a + 2^{1/2}b$.

3. Prove that addition is (a) commutative and (b) associative in S.

4. Prove that multiplication is (a) commutative and (b) associative in S.

5. Prove that multiplication is distributive over addition in S.

6. Let T be the set of all numbers of the form $a + 3^{1/2}b$, where a and b are rational numbers.

(a) Show that T is closed under addition and multiplication.
(b) Show that addition is commutative and associative in T.
(c) Show that multiplication is commutative and associative in T.
(d) Show that multiplication is distributive over addition in T.
(e) Show that there are identity elements for both addition and multiplication in T.

(f) Show that for each element in T there is an additive inverse which is also in T.

(g) Show that for each non-zero element in T there is a multiplicative inverse which is also in T.

(h) Does T form a field with the operations of addition and multiplication?

7. Prove that the set T (of Exercise 6) is closed under division.

8. Construct the addition and multiplication tables for the congruence classes modulo 6. Which class acts as an identity element for addition? Show the additive inverse element for each element in the set of congruence classes. Which class acts as an identity element for multiplication? Show all pairs of multiplicative inverses among the "non-zero" congruence classes. Do the set of congruence classes modulo 6 form a field with addition and multiplication?

9. Do as in Exercise 8 but using 7 as the modulus instead of 6.

10. Construct the addition and multiplication tables for the congruence classes modulo: (a) 2; (b) 3; (c) 4. Which, if any, of these three sets of congruence classes form fields with addition and multiplication?

11. What condition must a positive integer m satisfy in order that the set of congruence classes modulo m shall form a field with addition and multiplication? Give reasons in justification for your answer.

12. In ordinary arithmetic the statement $p \cdot q = 0$ can be true if, and only if, either $p = 0$ or $q = 0$ (or both). Does a similar principle apply in connection with the statement $[p] \cdot [q] = [0]$ concerning congruence classes? Justify your answer.

13. Construct the addition and multiplication tables for the subset $\{[2], [6], [10]\}$ of the set of congruence classes modulo 12.

CHAPTER 1.6

SUMMARY

SINCE the earliest beginnings of organized human life, men (and their women) have concerned themselves with the acquisition, possession, retention and trade of sets of things—such as clams, fishes, stones, clubs, furs, sheep, oil-wells, nuclear bombers and intercontinental missiles. Comparisons of such sets and the consequent knowledge of their relative extent have always seemed important. To facilitate an effective awareness of these things, men have cultivated certain fixed imagination patterns called **numbers.** Associated with each finite (or limited) set, men have conjured up in their minds a corresponding characteristic (called the **natural number** of elements in the set) which defines the capacity or incapacity of that set for being put into one-to-one correspondence with other sets. To determine which particular number should be associated with a given set, a procedure called **counting** was developed.

When two (or more) disjoint sets were combined into a union, the extension of counting to the union set gave rise to the **binary operation** of **addition.** Briefly, if the number of elements in one set is known to be a number x and the number of elements in a second (and disjoint) set is known to be a number y, then the "addition of x and y" is the process of finding the number of elements in the union of the two sets, this required number being represented symbolically by "$x + y$", the symbol " $+$ " representing the operation of addition. The natural numbers are **closed** under this operation, which means that for any two natural numbers x and y the resulting sum $x + y$, obtained by combining x and y by means of addition, is also a natural number. Addition of the natural numbers is **commutative,** which means that for any two natural numbers x and y,

$$x + y = y + x.$$

Addition of the natural numbers is also **associative,** which means that for any three natural numbers x, y and z,

$$x + (y + z) = (x + y) + z.$$

The number which corresponds to an empty set (that is, a set with no elements) is called **zero** and is represented by the symbol 0. For any natural number x it is true that

$$x + 0 = x \text{ and } 0 + x = x$$

and for this reason zero is called the **identity element for addition** for the system of natural numbers. Zero and the natural numbers are called, collectively, the **counting numbers** (or, by some people, the **whole numbers**).

Extended sums obtained by adding the same number repeatedly are called **multiples** or **products** and the process of forming such sums is called **multiplication.** Multiplication is also a binary (or two-number) operation since it makes use of two numbers (say x and y) to form a new number, written as $x \cdot y$, which is the sum of x ys. **The counting numbers are closed under multiplication and this operation is both commutative and associative for all counting numbers.** It is also **distributive over addition** for the counting numbers, which means that for any three counting numbers x, y and z,

$$x \cdot (y + z) = x \cdot y + x \cdot z$$

and

$$(y + z) \cdot x = y \cdot x + z \cdot x,$$

where the dot $\cdot$ is the symbol for multiplication. The counting number corresponding to a set containing a single element is called **one,** is represented by the symbol 1, and is the **identity**

element for multiplication in the counting numbers, which means that for any counting number x,

$$1 \cdot x = x \text{ and } x \cdot 1 = x.$$

Extended products obtained by multiplying repeatedly by the same number are called **powers** and the process of forming such products is called **exponentiation.** This operation is also a binary operation since it makes use of two numbers (say x and y) to form a new number which is the product of x ys. The counting numbers are closed under exponentiation (provided the base and exponent are not both zero). **Exponentiation is not commutative or associative** but **it is one-sidedly distributive over multiplication for all counting numbers.** The power in which the product of x ys is formed is represented by the symbol "y^x". The distributivity of exponentiation over multiplication is shown by the relation

$$(x \cdot y)^z = x^z \cdot y^z,$$

where x, y and z may represent any counting numbers (provided neither x and z nor y and z are both zero). **There is no identity element for exponentiation.**

The **inverse** of an operation is a second operation that undoes the effect of the first one.

The inverse of adding a number y to a number x to get a number z as a sum is the **subtraction** (or taking away) of y from z to get the original number x. Looked at from another viewpoint, the subtraction of a given number y from a given number z is the process of finding that number x to which y should be added to yield z as the resulting sum. The required number is called a **difference** and is represented by the symbol "$z - y$". **The counting numbers are not closed with respect to subtraction, subtraction is neither commutative nor associative, and there is no identity element for subtraction.** However, **multiplication is distributive over subtraction** which means that for any three counting numbers x, y and z,

$$x \cdot (y - z) = x \cdot y - x \cdot z \text{ and } (y - z) \cdot x = y \cdot x - z \cdot x.$$

The number system may be enlarged by adjoining to the system of counting numbers the **negative integers,** the latter being similar to the natural numbers but "opposite" to them in a certain sense. The new system is called the system of **integers** (the old natural numbers now being called the **positive integers**). Unlike the system of counting numbers, **the system of integers is closed under subtraction** so that for any two integers x and y the difference $x - y$ is also an integer. Of course, it is also closed under addition and multiplication.

The inverse of multiplying a given number x by a given number y to get a number z as a product is to determine what number x should be multiplied by a given number y to get a given number z as product. This inverse process is called **division** and the required number is called a **quotient** and is represented symbolically as

$$\frac{z}{y},\ z/y \ \text{or}\ z \div y.$$

The integers are not closed under division, division is neither commutative nor associative, and there is no identity element for division. Division is one-sidedly distributive over both addition and subtraction in the sense that, for any three integers x, y and z with z different from zero,

$$(x + y)/z = (x/z) + (y/z), \text{ or } \frac{x + y}{z} = \frac{x}{z} + \frac{y}{z},$$

and

$$(x - y)/z = (x/z) - (y/z), \text{ or } \frac{x - y}{z} = \frac{x}{z} - \frac{y}{z}.$$

The concepts of division and quotient can be extended and a new number system can be constructed which is closed under division by including, along with the integers, "numbers" of a new kind called **fractions.** If the integer one be interpreted as a unit of length (or some other partitionable type of unit) and if this one unit be partitioned into exactly x equal parts, then the fractional number associated with one of these parts is represented by the symbol $\frac{1}{x}$ (or $1/x$) and the sum of y of these numbers is the multiple $y \cdot \frac{1}{x}$ which is also written as $\frac{y}{x}$ (or y/x). The fraction y/x is also the result of partitioning a total of y units into x equal parts (that is, y/x represents one of these equal parts). The totality of all integers and fractions is called the **rational number system.** This is the simplest number system which is closed under all four of the operations of addition, subtraction, multiplication and division (all divisions being subject to the restriction that the divisor not be zero).

Since addition is commutative and $x + y$ and $y + x$ both have the same meaning as a number, the operation of finding a number x such that $x + y = z$ when y and z are given is no different than that of finding the number y such that $x + y = z$ when x and z and given. That is, **there is just one inverse operation for addition.** Similarly, since multiplication is also commutative, **there is just one inverse operation for multiplication.** However, since exponentiation is not commutative and x^y and y^x have different numerical meanings the inverse operation of finding a number x such that $x^y = z$ when y and z are given is distinct from the inverse operation of finding a number y such that $x^y = z$ when x and z are given; hence **there are two inverse operations for exponentiation.** The first of these inverse operations is called **root extraction** and the required number in this case is called the "yth root of z" and is represented by the symbol $z^{1/y}$. The second inverse operation is called **anti-exponentiation,** the required number in this case is called "the **logarithm** of z to the base x" and is represented symbolically as $\log_x z$. That is, if

$$x^y = z$$

then

$$x = z^{1/y} \text{ and } y = \log_x z.$$

Every rational number is representable either as a decimal fraction (that is, a "terminating" decimal) or as a non-terminating decimal. In the latter case what is actually involved is an infinite sequence of decimal fractions which converge to the represented number as a limit. Certain geometrically constructible quantities (such as $2^{1/2}$, $3^{1/2}$, etc.) which are not rational numbers are also representable as limits of convergent infinite sequences of rational numbers. Since all rational numbers can be represented by points on a straight line "number scale" and since any point on such a line can be represented as the limit of an infinite sequence of such "rational points" on the line it is possible to construct **a new number system which contains all the rational numbers and which is also closed under the operation of convergence to a limit** by defining a number to correspond to each point on a straight line. (The correspondence is completely determined by the selection of two points on the line—one to correspond to the number zero and the other to correspond to the number one.) The new number system is called the system of **real numbers.**

When the real numbers are further augmented by the adjunction of the number $(-1)^{1/2}$ (which is not to be found in the real number system) and all combinations of the form $a + bi$ (where a and b are any real numbers and $i = (-1)^{1/2}$) the resulting system, called **the complex number system, is found to be completely closed under addition, subtraction, multiplication, division, exponentiation, root extraction, anti-exponentiation and convergence to a limit.** As a bonus **it is also closed with respect to the solution of algebraic equations of any degree.** For these reasons it meets almost all needs encountered in the course of either practical or theoretical computation and may be considered as a nearly perfect number system. At least it is probably as nearly perfect a number system as we or even our distant descendants are likely to know.

If e is an identity element for an operation Δ in a set S, then an **inverse element** for Δ for an element s in S is defined as an element t such that $s \Delta t = e$ and $t \Delta s = e$.

We use the term **abstract system** to mean a set of elements on which one or more binary operations is defined.

Abstract systems are distinguished one from another by virtue of their possession of such properties as closure, associativity, commutativity, existence of identity elements, existence of inverse elements and distributivity for certain operations.

A **groupoid** is a non-empty set *A* of elements which is **closed** for a binary operation ① defined on *A*.

A **semi-group** is a groupoid *A* whose operation ① is **associative** for all elements in *A*.

A **monoid** is a semi-group *A* which possesses an **identity element** for its operation ①.

A **group** is a monoid *A* with binary operation ① such that for each element in *A* the **inverse** of that element for ① is also in *A*.

A **module** is a group *A* whose binary operation ① is **commutative** for all elements in *A*.

A **ring** is a module *A* with binary operation ① and on which a second binary operation ② is defined such that *A* is **closed** for ②, ② is **associative** for all elements in *A*, and ② is **distributive** over ① for all elements in *A*.

A **commutative ring** is a ring *A* with binary operations ① and ② in which ② is **commutative** for all elements in *A*.

A **ring with unity element** is a ring *A* with binary operations ① and ② such that *A* contains an **identity element for** ②.

An **integral domain** is a commutative ring *A* with unity element for binary operations ① and ② and such that if a and b are any elements in *A* for which a ② $b = 0$ (0 being the identity element for ① in *A*) then either $a = 0$ or $b = 0$.

A **field** is a commutative ring *A* with unity element for binary operations ① and ② and such that for each element in *A* (except the identity element for ①) the **inverse of that element for ② is also in *A*.**

The set of natural numbers forms a commutative semi-group with the binary operation of addition. It also forms a commutative monoid with multiplication. Multiplication is distributive over addition in this system.

The set of counting numbers forms a commutative monoid with each of the two binary operations of addition and multiplication.

Multiplication is, of course, distributive over addition in this system also.

The set of integers forms a module with addition and a commutative monoid with multiplication. The usual distributivity exists here and also distributivity of multiplication over subtraction.

The set of rational numbers form a field with the operations of addition and multiplication. So, of course, do the set of real numbers and also the set of complex numbers.

The purpose of the foregoing summary is, of course, to serve the needs of those who do not have the time to read the full text. For the benefit of those who are too busy even to read the summary the following summary of the summary is offered.

Exercises

1. One of your students asks: "What's the good of all this jazz about closure, commutativity, associativity, distributivity, identity elements, inverse elements, groups, fields and stuff?" Answer him.

2. Draw a diagram illustrating the inclusion relations between the various abstract systems.

3. Draw a diagram illustrating the inclusion relations between the various number systems.

4. Do numbers themselves exist in nature or are they man-made creations?

5. Give a different example for each of the following:

(a) a measurement that can be made using just the natural numbers (interpret the word "measurement" here to mean general numerical determination; this would include counting, of course);

(b) a measurement that cannot be made using just the natural numbers;

(c) a measurement that cannot be made using just the counting numbers;

(d) a measurement that cannot be made using just the integers;

(e) a measurement that cannot be made using just the rational numbers.

6. I have a specific number in mind which I would like you to guess. I will answer truthfully any "yes-or-no" type question that you ask me about the number. Design a sequence of twenty questions to be asked for the purpose of eliciting information about the number and possibly identifying it.

7. Let Q be the set of all numbers of the form $a + 2^{1/3}b$, where a and b are rational numbers. What abstract system, if any, does Q form with the operations of addition and multiplication?

8. Are there more rings than fields? Explain the reasoning on which your answer is based.

1.6.1. Summary of the Summary

1.6.1.1. *Properties of the number systems*

	Closure for									Commutativity and associativity for		Identity element for			Distributivity for		
	Addition	Subtraction	Multiplication	Division	Exponentiation (0^{-n} excl. for non-neg. n)	Root extraction	Anti-exponentiation	Convergence to a limit	Solution of algebraic equations	Addition and multiplication	Other operations	Addition	Multiplication	Other operations	Multiplication over add'n and subtr'n	Division over addition and subtraction	Exponentiation (and root extraction) over multiplication and division

Natural numbers	Yes	No	Yes	No	Yes	No	No	Not applicable	No	Yes	No	No	Yes	No	Yes	Yes (one-sided)	Yes (one-sided)
Counting numbers												Yes					
Integers		Yes			No												
Rational numbers				Yes				No									
Real numbers								Yes									
Complex numbers					Yes	Yes	Yes		Yes								

1.6.1.2. *The number systems as abstract systems*

<table>
<tr><th rowspan="2">Abstract system</th><th colspan="6">Number system</th></tr>
<tr><th>Natural numbers</th><th>Counting numbers</th><th>Integers</th><th>Rational numbers</th><th>Real numbers</th><th>Complex numbers</th></tr>
<tr><td>Groupoid</td><td colspan="6">For + ; also for ·</td></tr>
<tr><td>Semi-group</td><td colspan="6">For + ; also for ·</td></tr>
<tr><td>Monoid</td><td>For ·</td><td colspan="5">For + ; also for ·</td></tr>
<tr><td>Group</td><td></td><td></td><td>For +</td><td colspan="3">For + ; also for ·</td></tr>
<tr><td>Module</td><td></td><td></td><td>For +</td><td colspan="3">For + ; also for ·</td></tr>
<tr><td>Ring (commutative with unit element)</td><td></td><td></td><td colspan="4">For + and ·</td></tr>
<tr><td>Integral domain</td><td></td><td></td><td colspan="4">For + and ·</td></tr>
<tr><td>Field</td><td></td><td></td><td></td><td colspan="3">For + and ·</td></tr>
</table>

PART II

Computational Algorithms

CHAPTER 2.1

COMPUTATION WITH THE NATURAL NUMBERS

THE word algorithm (or algorism, as it is also spelled) means a standardized rule of procedure for carrying out certain mathematical computations, usually for a procedure which has been organized for maximum efficiency in getting certain desired numerical results.

A frequent concern throughout mathematical history has been the development of good algorithms (good from the point of view of being as simple and efficient as possible) first for the basic operations of addition, subtraction, multiplication and division and later for more intricate processes such as the extraction of second, third and higher roots, the solution of equations, and others. Since the search for such algorithms has been conducted by many people with varying cultural backgrounds at many different times and places, it is natural that many different algorithms may have evolved (even for a single operation such as multiplication), some of which may be better than others. However, since the evolutionary principle of survival of the fittest applies to ideas as well as to organisms, those algorithms which have survived the tests of time and competition to find their way into the schoolbooks of the past century have generally achieved this distinction by being the best of their kind that have been produced. The teaching of arithmetic—during the recent past, at least—has consisted largely of the teaching of these algorithms.

If textbook writers and teachers have not been wholly successful in arousing in their students an appreciation of the usefulness and other merits of these algorithms, it is only fair to look into the way in which the textbook writers have written, the way the teachers have taught and the way the students have studied before deciding that the algorithms should be discarded in favor of others that are not as good—or even, as has been seriously suggested in some quarters, in favor of pure trial-and-error methods.

It should be realized by all, of course, that no useful learning of computational algorithms—or anything else—can take place without serious concentration of attention and will to learn on the part of the student. Most textbook writers and teachers, at least, do recognize this. With such concentration and will, a student might learn much thru his own discoveries even if textbook writers and teachers did not exist. However, the rate of discovery and consequent scope of learning may be greatly increased if textbook writers and teachers take advantage of their opportunity to provide carefully planned **guidance** for the student's exploratory thinking in such a way as to optimize the probability of significant and fruitful discovery. This fact is the basis on which the "discovery" method of teaching legitimately rests. It should not be overlooked—but frequently is—that without such carefully planned guidance any attempt to employ discovery-type learning methods degenerates into the old-fashioned "sink-or-swim" method of which we have already had too much.

Along with the provision of systematic guidance directed toward fruitful discovery another contribution toward increasing the rapidity and scope of learning which can be provided by textbook writers and teachers is that of **skillful explanation.** Indeed, since thru this device (as history confirms) the process of learning may be accelerated and its scope enlarged more powerfully than thru any other method, the claim that this is the first duty of textbook writers and teachers should need no defense.

The sometimes expressed fear that students will be bored due to lack of challenge if "everything is explained to them" will have no basis in fact if explanation is given on those things which students cannot be expected to discover for themselves—or which for one reason or another they somehow do not manage to discover for themselves—and if the explanations are given in a genuinely skillful, authoritative and interesting manner. Of course, students should be invited, encouraged and guided to discover for themselves as much as they can, but in view of the fact that it has quite literally taken ages for each new concept in mathematics to be discovered (or invented, if you prefer that point of view) the shortness of life makes it immediately obvious that we simply cannot afford to wait until students discover quite independently—without explanation, guidance or other help from textbook or teacher—all the concepts and facts which could be of value to them during their lives.

The remainder of this book consists of explanations of how the common algorithms of arithmetic are derived from considerations of the structural nature of the number systems of arithmetic. The objective of these explanations is to enable teachers to give both correctly informative and satisfying answers to the natural questions that will occur to their students as these algorithms are presented to them.

2.1.1. The Addition Algorithm

We shall start with some very simple examples in addition and, by working our way thru progressively more complicated ones, develop simultaneously both the need for, and the nature of, our final algorithm.

In adding three to five we may either obtain the result as the third successor of five (by extended counting) or we may use the addition table on page 35 to obtain

$$5 + 3 = 8.$$

In adding seven to five we may use either of the two methods mentioned in the preceding problem but we must also make use of the fact that the result obtained (twelve) is also equal to ten plus two. (This fact, of course, is implicit in the use of the base-ten positional notation used in the addition table.) Justification for this equality lies in the associative law thus:

$$5 + 7 = 5 + (5 + 2) = (5 + 5) + 2 = 10 + 2 = 12.$$

In adding 27 to 65 we proceed as follows.

$65 + 27 = (6 \cdot 10 + 5) + (2 \cdot 10 + 7)$, by nature of positional notation,
$= (6 \cdot 10 + 2 \cdot 10) + (5 + 7)$, by associativity and commutativity,
$= (6 + 2) \cdot 10 + (5 + 7)$, by distributivity,
$= 8 \cdot 10 + (1 \cdot 10 + 2)$, by the addition table,
$= (8 \cdot 10 + 1 \cdot 10) + 2$, by associativity,
$= (8 + 1) \cdot 10 + 2$, by distributivity,
$= 9 \cdot 10 + 2$, by the addition table,
$= 92$, by the principle of positional notation.

The combination here of the $6 \cdot 10$ (six tens) and the $2 \cdot 10$ (two tens) to obtain the $8 \cdot 10$ (eight tens) is straightforward, as is the combination of five and seven to obtain twelve. The new maneuver which has been introduced into the computation is the separation of the ten from the two in the twelve, the transfer (or carrying over) of this ten from its original association with the two to a new association with the eight tens, and its final combination with the eight tens to form nine tens. This sort of transfer operation is commonly called **carrying** and it is said (in this case) that we have "carried a one (meaning *one ten*) from the units into the tens". While carrying may sometimes not be necessary in addition problems, it is frequently necessary in most of them and its nature should be well understood. It will be illustrated further in the following examples.

In adding 87 to 65 two "carries" may be seen to be involved in the following procedure.

$$\begin{aligned} 65 + 87 &= (6 \cdot 10 + 5) + (8 \cdot 10 + 7) \\ &= (6 \cdot 10 + 8 \cdot 10) + (5 + 7) \\ &= 14 \cdot 10 + 12 \\ &= (10 + 4) \cdot 10 + (10 + 2) \\ &= (10 \cdot 10 + 4 \cdot 10) + (1 \cdot 10 + 2) \\ &= 10^2 + (4 \cdot 10 + 1 \cdot 10) + 2 \\ &= 1 \cdot 10^2 + 5 \cdot 10 + 2 \\ &= 152. \end{aligned}$$

Here a one is carried into the tens and another one is carried into the hundreds.

In forming the sum $65 + 87 + 79$ we first write (or think)

$$\begin{aligned} 65 + 87 + 79 &= (6 \cdot 10 + 5) + (8 \cdot 10 + 7) + (7 \cdot 10 + 9) \\ &= (6 \cdot 10 + 8 \cdot 10 + 7 \cdot 10) + (5 + 7 + 9). \end{aligned}$$

After this we could proceed in different ways. In simplifying the sum $(5 + 7 + 9)$ we might first convert $7 + 9$ into the form 16, or $1 \cdot 10 + 6$, carry the one ten, then add the remaining 6 to 5 to get 11, or $1 \cdot 10 + 1$, and then carry this second ten. In practice, however, it is found more efficient and convenient to complete the addition of the 5 and 16 thus:

$$\begin{aligned} 5 + 7 + 9 &= 5 + 16 \\ &= 5 + (1 \cdot 10 + 6) \\ &= 1 \cdot 10 + (5 + 6) \\ &= 1 \cdot 10 + (1 \cdot 10 + 1) \\ &= (1 \cdot 10 + 1 \cdot 10) + 1 \\ &= 2 \cdot 10 + 1, \end{aligned}$$

and then carry the two tens at one time. Similarly, we simplify the sum $6 \cdot 10 + 8 \cdot 10 + 7 \cdot 10$ thus:

$$\begin{aligned} 6 \cdot 10 + 8 \cdot 10 + 7 \cdot 10 &= (6 + 8 + 7) \cdot 10 \\ &= (6 + 15) \cdot 10 \\ &= 21 \cdot 10 \\ &= (2 \cdot 10 + 1) \cdot 10 \\ &= 2 \cdot 10^2 + 1 \cdot 10; \end{aligned}$$

then finally,

$$\begin{aligned} 65 + 87 + 79 &= (6 \cdot 10 + 8 \cdot 10 + 7 \cdot 10) + (5 + 7 + 9) \\ &= (2 \cdot 10^2 + 1 \cdot 10) + (2 \cdot 10 + 1) \\ &= 2 \cdot 10^2 + (1 \cdot 10 + 2 \cdot 10) + 1 \\ &= 2 \cdot 10^2 + 3 \cdot 10 + 1 \\ &= 231. \end{aligned}$$

Notice that in this entire process there has been a carrying of two tens (that is, a carrying of "two from units into tens") and also a carrying of two hundreds ("two from tens into hundreds").

Up to this point we have been concerned only with procedures which are consistent with the known properties of the natural numbers, and we have carried out these procedures in a written form known as the "continued equation form". While these procedures (and the form) have been logically correct, there is no denying that they are cumbersome and would become increasingly so for addition problems involving either large summands or a large number of summands. For such problems we now seek a more convenient and expeditious way of handling the necessary details of the computation without diverting time and energy to other details which may be ignored or taken for granted. The way of doing this that has so far found most widespread acceptance as "standard" is to arrange the summands in a vertical column with all the rightmost digits exactly aligned vertically thus:

$$\begin{array}{c} 6\,5 \\ 8\,7 \\ 7\,9 \\ \hline \end{array}$$

A horizontal line segment is drawn under the last summand to separate the summands from other numbers associated with them (including the sum) which are to be written later. The summands as written, of course, stand for

$$\begin{array}{c} 6 \cdot 10 + 5 \\ 8 \cdot 10 + 7 \\ 7 \cdot 10 + 9 \\ \hline \end{array}$$

and the preliminary additions shown previously in equation form can also be shown by

$$\begin{array}{c} 6 \cdot 10 + 5 \\ 8 \cdot 10 + 7 \\ 7 \cdot 10 + 9 \\ \hline (6 + 8 + 7) \cdot 10 + (5 + 7 + 9) \end{array}$$

or

$$\begin{array}{r} 6 \cdot 10 + 5 \\ 8 \cdot 10 + 7 \\ 7 \cdot 10 + 9 \\ \hline 21 \cdot 10 + 21 \end{array}$$

or, using the more compact positional notation (to conserve space and energy),

$$\begin{array}{r} 6\,5 \\ 8\,7 \\ 7\,9 \\ \hline 2\,1 \\ 2\,1\,0. \end{array}$$

There now remains only the combining of the sum (21) of the units digits with the sum ($21 \cdot 10$, or 210) of the tens, and these two numbers as last written are in a convenient position for making this final combination. Again adding each column of digits separately in the 21 and 210 we obtain

$$\begin{array}{r} 2 \cdot 10 + 1 \\ 2 \cdot 10^2 + 1 \cdot 10 + 0 \\ \hline 2 \cdot 10^2 + 3 \cdot 10 + 1 \end{array}$$

or

$$\begin{array}{r} 2\,1 \\ 2\,1\,0 \\ \hline 2\,3\,1, \end{array}$$

yielding 231 as the final form of the required sum.

The method is illustrated again in the following somewhat more complicated example involving the sum 478 + 962 + 4283 + 53 + 1902.

```
  478
  962
 4283
   53
 1902
 ----
   18 ← (Sum of units)
  260 ← (Sum of tens)
 2400 ← (Sum of hundreds)
 5000 ← (Sum of thousands)
```

The 18 written here is, of course, the sum $8 + 2 + 3 + 3 + 2$ of the units digits; the 260 is the sum $(7 + 6 + 8 + 5 + 0) \cdot 10$ of all the tens; the 2400 is the sum $(4 + 9 + 2 + 9) \cdot 10^2$ of all the hundreds; and the 5000 is the sum $(4 + 1) \cdot 10^3$ of all the thousands. Now treating the 18 as $1 \cdot 10 + 8$, the 260 as $2 \cdot 10^2 + 6 \cdot 10$, the 2400 as $2 \cdot 10^3 + 4 \cdot 10^2$ and the 5000 as $5 \cdot 10^3$ we combine all these preliminary sums to get

$$(2 + 5) \cdot 10^3 + (2 + 4) \cdot 10^2 + (1 + 6) \cdot 10 + 8$$

or

$$7 \cdot 10^3 + 6 \cdot 10^2 + 7 \cdot 10 + 8$$

or

$$7678$$

as the final required sum, the entire computational writing appearing as

```
  478
  962
 4283
   53
 1902
 ----
   18
  260
 2400
 5000
 ----
 7678.
```

This is obviously much shorter than the same story would be when told in equation form. However, it can be made still shorter. If each of the digits which is to be carried (the 1 standing for one ten in the 18, the 2 standing for two hundreds in the 260, and the 2 standing for two thousands in the 2400) is written small and all the zeros are omitted, then all of the preliminary sums can be written on the same line thus:

$$\begin{array}{r} 478 \\ 962 \\ 4283 \\ 53 \\ 1902 \\ \hline 5_2 4_2 6_1 8. \end{array}$$

If now the carried digits are added to the other digits occupying the same spaces into which the carried digits are carried, then the final form of the required sum is obtained as 7678. A further refinement which may be made here is to write the small carried digits one line higher and then perform the last described addition of these carried digits mentally as each column of digits is added, writing down just the final sum in the space below the line thus:

$$\begin{array}{r} 478 \\ 962 \\ 4283 \\ 53 \\ 1_2 9_2 0_1 2 \\ \hline 7678. \end{array}$$

This compact form is one which average children in even the lower elementary grades have been known to master without undue difficulty when it was presented to them with appropriate explanation. For older children a still further refinement may be added by having them keep track mentally of all carrying information and write just

$$\begin{array}{r} 478 \\ 962 \\ 4283 \\ 53 \\ 1902 \\ \hline 7678. \end{array}$$

Needless (?) to say, when students are assigned practical problems in addition, a great deal more attention should be devoted to verifying that they can demonstrate their understanding of the significance of the column-by-column addition of digits and the subsequent carrying than to a mere "checking" of the "answer". [Incidentally, it might be well to suggest here and now that the use of the word "answer" be henceforth abandoned in all mathematical work except insofar as it is used to mean "a reply to an interrogation". When a student is asked what the specific (mathematical) objective is in a particular problem, he should be aware and able to state that it is "to find a certain sum" whenever this is the case, or "to compute a certain square root" if this be the case, and so on. When the reply to this sort of question is "To get the answer", this is a pretty good indication that the student has not learned the difference between a sum and a square root and hasn't the faintest idea what the problem is about. When students are permitted (or even encouraged) to use the word "answer" to mean everything from "sum" to "square root" and beyond, you can't blame them too much for the resulting confusion.]

Exercises

1. (a) State briefly whatever historical information you may have (or your best conjecture, if you do not have actual information) concerning the time required by mankind to develop the concept of the operation of addition and a satisfactory practical algorithm to handle the accompanying details of computation. (b) Giving careful consideration to all the other things which the average child is required to learn during his school years, estimate the total time which is available to him for the purpose of learning about the concept of addition and one or more related algorithms. (c) To what extent do you believe it workable to put an average school-child "on his own" and rely on his ability to learn about the concept of addition and a good algorithm for addition solely on the basis of his own unaided discoveries?

2. Describe what you consider to be a legitimate, intelligent and productive version of the "discovery method" of teaching.

3. (a) Perform the indicated addition in the expression 264 + 78 + 939, showing all logical steps in the continued equation form. [Note: "All logical steps" implies all applications of associative, commutative and distributive laws, addition facts, etc.] (b) Perform the same addition, using only the most condensed algorithm.

4. (a) Using the condensed algorithm, add the numbers 6,264,591, 2,467,999, 5,483,963, 9,402,211, 8,260,439, 7,787,922, 1,190,929, 592,004 and 26,378. (b) Estimate the number of pages that you might use in showing all the details of this same computation in "continued equation form".

5. Interpreting all numbers as in octal notation, add the following.

36,742
25,366
43,212
7,627
746,513
777
44,567

6. Interpreting all numbers as in binary notation, add the following.

1,011
11,100
111
101,110
10,101

7. (a) Formulate a problem in addition which you think has excellent illustrative properties and hence also excellent instructional value. (b) Point out the nature of the particular illustrative properties which your problem has.

8. The addition algorithm has been described in connection with specific examples of addition. State the algorithm in words without reference to any particular numerical example.

9. Think carefully of all of your own learning experiences in arithmetic and estimate as accurately as you can: (a) the percentage of your present arithmetic knowledge which you learned thru explanation given by teachers or other persons; (b) the percentage which you learned thru reading and study of books, working of assigned problems, etc.; (c) the percentage which you learned thru your own unguided investigations and unaided discoveries.

2.1.2. The Multiplication Algorithm

In multiplying 4627 by 538 the application of the distributive law and the other natural laws leads us immediately to the following logical and accurate—altho somewhat cumbersome—evaluation of the product.

$$538 \cdot 4627 = (5 \cdot 10^2 + 3 \cdot 10 + 8) \cdot (4 \cdot 10^3 + 6 \cdot 10^2 + 2 \cdot 10 + 7)$$

$$\begin{aligned} &= 5 \cdot 10^2 \cdot (4 \cdot 10^3 + 6 \cdot 10^2 + 2 \cdot 10 + 7) \\ &\quad + 3 \cdot 10 \cdot (4 \cdot 10^3 + 6 \cdot 10^2 + 2 \cdot 10 + 7) \\ &\quad\quad + 8 \cdot (4 \cdot 10^3 + 6 \cdot 10^2 + 2 \cdot 10 + 7) \end{aligned}$$

$$\begin{aligned} &= 5 \cdot 10^2 \cdot 4 \cdot 10^3 + 5 \cdot 10^2 \cdot 6 \cdot 10^2 + 5 \cdot 10^2 \cdot 2 \cdot 10 + 5 \cdot 10^2 \cdot 7 \\ &\quad + 3 \cdot 10 \cdot 4 \cdot 10^3 + 3 \cdot 10 \cdot 6 \cdot 10^2 + 3 \cdot 10 \cdot 2 \cdot 10 + 3 \cdot 10 \cdot 7 \\ &\quad\quad + 8 \cdot 4 \cdot 10^3 + 8 \cdot 6 \cdot 10^2 + 8 \cdot 2 \cdot 10 + 8 \cdot 7 \end{aligned}$$

$$\begin{aligned} &= 20 \cdot 10^5 + 30 \cdot 10^4 + 10 \cdot 10^3 + 35 \cdot 10^2 \\ &\quad + 12 \cdot 10^4 + 18 \cdot 10^3 + 6 \cdot 10^2 + 21 \cdot 10 \\ &\quad\quad + 32 \cdot 10^3 + 48 \cdot 10^2 + 16 \cdot 10 + 56. \end{aligned}$$

It will be noticed here that the positioning of the various partial-product terms is not haphazard or coincidental, but has been arranged systematically to keep multiples of like powers of ten aligned in vertical columns in order to facilitate their combination. Thus, after making appropriate application of the associative and commutative laws of addition, and the distributive law, we have

$$\begin{aligned} 538 \cdot 4627 &= 20 \cdot 10^5 + (30 + 12) \cdot 10^4 + (10 + 18 + 32) \cdot 10^3 \\ &\quad + (35 + 6 + 48) \cdot 10^2 + (21 + 16) \cdot 10 + 56 \\ &= 20 \cdot 10^5 + 42 \cdot 10^4 + 60 \cdot 10^3 + 89 \cdot 10^2 + 37 \cdot 10 + 56. \end{aligned}$$

Now using the carrying principle, we rewrite the result as

$$
\begin{aligned}
538 \cdot 4627 &= 2 \cdot 10^6 + (4 \cdot 10^5 + 2 \cdot 10^4) + 6 \cdot 10^4 + (8 \cdot 10^3 \\
&\quad + 9 \cdot 10^2) + (3 \cdot 10^2 + 7 \cdot 10) + (5 \cdot 10 + 6) \\
&= 2 \cdot 10^6 + 4 \cdot 10^5 + (2 \cdot 10^4 + 6 \cdot 10^4) + 8 \cdot 10^3 \\
&\quad + (9 \cdot 10^2 + 3 \cdot 10^2) + (7 \cdot 10 + 5 \cdot 10) + 6 \\
&= 2 \cdot 10^6 + 4 \cdot 10^5 + 8 \cdot 10^4 + 8 \cdot 10^3 + 12 \cdot 10^2 \\
&\quad + 12 \cdot 10 + 6 \\
&= 2 \cdot 10^6 + 4 \cdot 10^5 + 8 \cdot 10^4 + 8 \cdot 10^3 + (1 \cdot 10^3 \\
&\quad + 2 \cdot 10^2) + (1 \cdot 10^2 + 2 \cdot 10) + 6 \\
&= 2 \cdot 10^6 + 4 \cdot 10^5 + 8 \cdot 10^4 + (8 \cdot 10^3 + 1 \cdot 10^3) \\
&\quad + (2 \cdot 10^2 + 1 \cdot 10^2) + 2 \cdot 10 + 6 \\
&= 2 \cdot 10^6 + 4 \cdot 10^5 + 8 \cdot 10^4 + 9 \cdot 10^3 + 3 \cdot 10^2 \\
&\quad + 2 \cdot 10 + 6 \\
&= 2{,}489{,}326.
\end{aligned}
$$

We do not claim that this procedure is brief, convenient or an efficient way to obtain the result—but only that it is logical and reliably accurate. To obtain the desired brevity, convenience and efficiency we must find a way to preserve all the essential computations (multiplications and additions) while omitting the writing of all notational details which are not essential to the final result. A step in this direction is to rewrite all the partial products in standard positional notation rather than as multiples of powers of ten thus:

$$
\begin{aligned}
538 \cdot 4627 = 2{,}000{,}000 &+ 300{,}000 + 10{,}000 + 3500 \\
&+ 120{,}000 + 18{,}000 + 600 + 210 \\
&+ 32{,}000 + 4800 + 160 + 56.
\end{aligned}
$$

We may now eliminate the operational symbols $\cdot$ and $+$ and the equality symbol $=$ by writing all numbers in column form with units vertically in line, tens vertically aligned immediately to the left of the units, hundreds aligned immediately to the left of the tens, and so on, thus:

```
   4627
    538
-------
2000000
 300000
  10000
   3500
 120000
  18000
    600
    210
  32000
   4800
    160
     56,
```

with a horizontal line segment drawn to separate the multiplicand and multiplier from the various partial products. Next, by maintaining the columnar arrangement and applying the principle of positional notation we may omit all zeros and represent the above computational record by writing

```
   4627
    538
-------
2
 3
  1
   35
 12
  18
    6
    21
  32
   48
    16
     56.
```

The entries 2, 3, 1 and 35 (standing for $2 \cdot 10^6 + 3 \cdot 10^5 + 1 \cdot 10^4 + 3 \cdot 10^3 + 5 \cdot 10^2$, the product obtained by multiplying 4627 by $5 \cdot 10^2$) can now be compacted into a single line and written as 23135 in a position to represent 2,313,500. Similarly, the entries 12, 18, 6 and 21 can be written as $12_1 86_2 1$ where the small numerals have the usual "carrying" significance; and the entries 32, 48, 16 and 56 can be written as $32_4 8_1 6_5 6$. The computational record now becomes

$$\begin{array}{r} 4\,6\,2\,7 \\ 5\,3\,8 \\ \hline 2\,3\,1\,3\,5 \\ 1\,2_1 8\,6_2 1 \\ 3\,2_4 8_1 6_5 6, \end{array}$$

where the last three lines represent the respective products $500 \cdot 4627$, $30 \cdot 4627$ and $8 \cdot 4627$. If the carrying be done mentally as each of these products is computed and recorded, the record then becomes

$$\begin{array}{r} 4\,6\,2\,7 \\ 5\,3\,8 \\ \hline 2\,3\,1\,3\,5 \\ 1\,3\,8\,8\,1 \\ 3\,7\,0\,1\,6 \end{array}$$

which, of course, stands for

$$\begin{array}{r} 4\,6\,2\,7 \\ 5\,3\,8 \\ \hline 2\,3\,1\,3\,5\,0\,0 \\ 1\,3\,8\,8\,1\,0 \\ 3\,7\,0\,1\,6. \end{array}$$

The three partial products 2,313,500, 138,810 and 37,016 are now added column-by-column to obtain the final product thus:

```
    4627
     538
--------
 2313500
  138810
   37016
--------
2,489,326
```

In general practice, the zeros are not written. Also, since it is customary to multiply the multiplicand (4627) first by the number (8) of units in the multiplier, then multiply it by the number (3) of tens, and then by the number (5) of hundreds instead of in the reverse order as we have done, the standard presentation of the computational record for this particular problem is

```
    4627
     538
--------
   37016
  13881
 23135
--------
2,489,326.
```

By the application of still further mental work, the computational record can be reduced to the "one-line" form

```
    4627
     538
--------
2,489,326.
```

or

$$538 \cdot 4627 = 2{,}489{,}326$$

but this feat is not recommended for presentation to all students and it is suggested that it be reserved for those prodigies who demand "greater challenge" and "enrichment".

2.1.2.1. *Alternative explanation of the multiplication algorithm*

An alternative way in which to approach the development of the multiplication algorithm is to consider first the problem of multiplication by a single digit. For instance, in multiplying 4627 by 8, we first consider the formal relation

$$\begin{aligned} 8 \cdot 4627 &= 8 \cdot (4 \cdot 10^3 + 6 \cdot 10^2 + 2 \cdot 10 + 7) \\ &= 8 \cdot (4 \cdot 10^3) + 8 \cdot (6 \cdot 10^2) + 8 \cdot (2 \cdot 10) + 8 \cdot 7 \\ &= (8 \cdot 4) \cdot 10^3 + (8 \cdot 6) \cdot 10^2 + (8 \cdot 2) \cdot 10 + 8 \cdot 7 \\ &= 32 \cdot 10^3 + 48 \cdot 10^2 + 16 \cdot 10 + 56 \\ &= 32{,}000 + 4800 + 160 + 56. \end{aligned}$$

The latter sum may be added in the columnar manner thus:

$$\begin{array}{r} 3\,2\,0\,0\,0 \\ 4\,8\,0\,0 \\ 1\,6\,0 \\ 5\,6 \\ \hline 3\,7{,}0\,1\,6; \end{array}$$

or, by the positional notation principle, the nature of the quantities involved (thousands, hundreds, tens, units, etc.) may be shown by the positioning of their multipliers thus:

$$\begin{array}{lllll} 3 & 2 & & & \\ & 4 & 8 & & \\ & & 1 & 6 & \\ & & & 5 & 6 \\ \hline 3 & 7, & 0 & 1 & 6. \end{array}$$

The notations representing the partial products may be further condensed into a single line by writing them as follows:

$$3\;2_4 8_1 6_5 6,$$

whereupon, by adding the small "carried" digits, we obtain the result

$$3\ 7{,}0\ 1\ 6.$$

Of course, it is also possible, instead of writing the carried digits as indicated, to carry and add them mentally, thus producing directly the result

$$8 \cdot 4627 = 37{,}016.$$

In multiplying 4627 by 30, the procedure is essentially the same except for a factor of ten which may be handled as follows.

$$\begin{aligned} 30 \cdot 4627 &= (3 \cdot 10) \cdot 4627 \\ &= 3 \cdot (10 \cdot 4627) \\ &= 3 \cdot (4627 \cdot 10) \\ &= (3 \cdot 4627) \cdot 10 \\ &= 13{,}881 \cdot 10 \\ &= 138{,}810. \end{aligned}$$

Similarly,

$$\begin{aligned} 500 \cdot 4627 &= (5 \cdot 10^2) \cdot 4627 \\ &= 5 \cdot (10^2 \cdot 4627) \\ &= 5 \cdot (4627 \cdot 10^2) \\ &= (5 \cdot 4627) \cdot 10^2 \\ &= 23{,}135 \cdot 10^2 \\ &= 2{,}313{,}500. \end{aligned}$$

Consequently, in evaluating the product $538 \cdot 4627$, we first express it as the sum of the above three partial products thus:

$$\begin{aligned} 538 \cdot 4627 &= (5 \cdot 10^2 + 3 \cdot 10 + 8) \cdot 4627 \\ &= (5 \cdot 10^2) \cdot 4627 + (3 \cdot 10) \cdot 4627 + 8 \cdot 4627 \\ &= (5 \cdot 4627) \cdot 10^2 + (3 \cdot 4627) \cdot 10 + 8 \cdot 4627 \\ &= 23{,}135 \cdot 10^2 + 13{,}881 \cdot 10 + 37{,}016. \end{aligned}$$

Then, in evaluating this sum, we may either list the terms fully in columnar form to add them thus:

```
   3 7 0 1 6
 1 3 8 8 1 0
2 3 1 3 5 0 0
-------------
2,4 8 9,3 2 6;
```

or, using the principle of positional notation,

```
    3 7 0 1 6
  1 3 8 8 1
2 3 1 3 5
-------------
2,4 8 9,3 2 6.
```

The complete record of the computation, showing multiplicand, multiplier, partial products and final product, all located with positional significance in a column, is, of course,

```
      4 6 2 7
        5 3 8
-------------
    3 7 0 1 6
  1 3 8 8 1
2 3 1 3 5
-------------
2,4 8 9,3 2 6.
```

Exercises

1. (a) Show in continued equation form all logical steps (that is, all applications of fundamental laws, facts of addition and multiplication, etc.) in the multiplication of 358 by 76. (b) Carry out the same multiplication by means of the standard algorithm.

2. Describe the standard multiplication algorithm without referring to any particular problem.

3. Explain in detail the mental steps used in "one-line" multiplication. (See bottom of page 184.)

4. (a) Multiply 29,653 by 300,102 according to the standard algorithm. (b) Using this problem as an example, show how the standard algorithm can be shortened whenever a zero occurs in the multiplier.

5. Interpreting all numbers as in the octal system, use the standard multiplication algorithm to find the following products: (a) 34·75; (b) 326·57; (c) 762·545; (d) 76,451·37,264.

6. Interpreting all numbers as in the binary system, use the standard multiplication algorithm to find the following products: (a) $11 \cdot 10$; (b) $101 \cdot 11$; (c) $10{,}111 \cdot 1{,}101$; (d) $1{,}101{,}011 \cdot 101{,}111$.

7. What is the base of the positional system in which $45 \cdot 32 = 2133$?

8. What percentage of the time allotted to the teaching of multiplication should an elementary teacher spend in showing his or her students methods of multiplication which are different from the conventional algorithm and perhaps interesting but which do not produce products as rapidly, directly or easily as the conventional algorithm? Why?

2.1.3. The Subtraction Algorithm

The operation of subtraction can be reduced to a simple algorithm even more easily than the operation of addition, for while most addition problems are problems in *extended* addition—that is, the addition of many summands—the ordinary subtraction problem involves a strictly binary operation.

The logic of subtracting 45 from 97 may be indicated in continued equation form as follows:

$$\begin{aligned} 97 - 45 &= (9 \cdot 10 + 7) - (4 \cdot 10 + 5) \\ &= 9 \cdot 10 + 7 - 4 \cdot 10 - 5 \\ &= (9 \cdot 10 - 4 \cdot 10) + (7 - 5) \\ &= (9 - 4)\,10 + (7 - 5) \\ &= 5 \cdot 10 + 2 \\ &= 52. \end{aligned}$$

As in the case of addition, the numbers may be arranged in a vertical column for convenience in subtracting units from units and tens from tens, and may be shown, with the resulting difference as follows:

$$\begin{array}{r} 9\,7 \\ 4\,5 \\ \hline 5\,2. \end{array}$$

In a case such as the one involving the subtraction of 45 from 93, an extra step in the reasoning is involved. Starting out as before, we write

$$\begin{aligned} 93 - 45 &= (9 \cdot 10 + 3) - (4 \cdot 10 + 5) \\ &= 9 \cdot 10 + 3 - 4 \cdot 10 - 5 \\ &= (9 \cdot 10 - 4 \cdot 10) + (3 - 5) \\ &= (9 - 4) \cdot 10 + (3 - 5) \\ &= 5 \cdot 10 + (3 - 5). \end{aligned}$$

However, since it is usually our desire in problems such as this to express all parts of the required difference as *positive* numbers, we prefer not to deal with the negative difference $(3 - 5)$. To avoid dealing with it we arrange our computational reasoning as follows (in what amounts to a reversal of the "carrying" operation used in addition).

$$\begin{aligned} 93 - 45 &= (9 \cdot 10 + 3) - (4 \cdot 10 + 5) \\ &= 9 \cdot 10 + 3 - 4 \cdot 10 - 5 \\ &= (8 + 1) \cdot 10 + 3 - 4 \cdot 10 - 5 \\ &= (8 \cdot 10 + 1 \cdot 10) + 3 - 4 \cdot 10 - 5 \\ &= 8 \cdot 10 + (1 \cdot 10 + 3) - 4 \cdot 10 - 5 \\ &= 8 \cdot 10 + 13 - 4 \cdot 10 - 5 \\ &= (8 \cdot 10 - 4 \cdot 10) + (13 - 5) \\ &= (8 - 4) \cdot 10 + (13 - 5) \\ &= 4 \cdot 10 + 8 \\ &= 48. \end{aligned}$$

The operation of removing one of the tens from the "tens column" in the 93 and combining it with the 3 (standing for three units) makes it possible to subtract the 5 (five units) in the 45 from the resulting 13 units and obtain the positive difference of 8 (eight units). This practice (which is common in subtraction) of removing one from a particular column and replacing it by ten in the adjacent right-hand column to make subtraction there possible (with a resulting positive difference) is called **borrowing.** (In this particular example we say that we "borrow a one from the tens column".)

When using the column algorithm with the above example the

borrowing may be indicated by crossing out the 9, writing an 8 above it, and writing a small 1 just to the left of the 3 thus:

$$\begin{array}{r} 8\phantom{{}_13} \\ \not{9}{}_13 \\ 4\;5 \\ \hline 4\;8. \end{array}$$

The 5 is then subtracted from the artificially obtained 13, the 4 is subtracted from the 8, and the resulting difference of 48 is written below the horizontal line.

For more advanced pupils the crossing out of the 9 may be omitted and its replacement by the 8 may be done mentally, rather than in writing, resulting in the following form:

$$\begin{array}{r} 9{}_13 \\ 4\;5 \\ \hline 4\;8. \end{array}$$

A still simpler form, suitable for relatively mature children, involves attending to all aspects of the borrowing operation mentally and writing just

$$\begin{array}{r} 9\;3 \\ 4\;5 \\ \hline 4\;8. \end{array}$$

We now illustrate the algorithm for a more complicated problem, both with and without the written notations for emphasizing the borrowing details.

$$\begin{array}{r} 2{}_11\;9{}_16{}_12\;4{}_12\;3 \\ 1\;7\;4\;8\;5\;2\;6\;1 \\ \hline 4{,}4\;7\;7{,}1\;6\;2 \end{array} \qquad \begin{array}{r} 2\;1\;9\;6\;2\;4\;2\;3 \\ 1\;7\;4\;8\;5\;2\;6\;1 \\ \hline 4{,}4\;7\;7{,}1\;6\;2 \end{array}$$

In deciding when to permit a particular child to take the short cuts of omitting the special auxiliary notations having to do with carrying and borrowing, there is no objection to encouraging this whenever the child himself feels ready to do so—provided, of course, that he can demonstrate his ability to handle the details mentally without making errors.

Exercises

1. (a) Show in continued equation form all logical steps in the subtraction of 486 from 723. (b) Show the same subtraction using the column algorithm.

2. In the following subtractions use the column algorithm, first showing all borrowing by appropriate notations; then rewrite the work in the simpler form, doing all borrowing mentally. (a) Subtract 25 from 87. (b) Subtract 27 from 85. (c) Subtract 264 from 813. (d) Subtract 42,596,213,487 from 819,422,654,257.

3. State the subtraction algorithm in words without reference to any particular numerical example.

4. Formulate a subtraction problem having good illustrative properties and point out what these properties are.

5. Interpreting all numbers as in octal notation, perform the following subtractions.

$$\text{(a)}\ \begin{array}{r} 75 \\ 43 \\ \hline \end{array} \qquad \text{(b)}\ \begin{array}{r} 7\,5 \\ 4\,7 \\ \hline \end{array} \qquad \text{(c)}\ \begin{array}{r} 2\,5\,6 \\ 1\,7\,7 \\ \hline \end{array} \qquad \text{(d)}\ \begin{array}{r} 4\,3\,5\,2 \\ 2\,7\,2\,6 \\ \hline \end{array}$$

6. Interpreting all numbers as in binary notation, perform the following subtractions.

$$\text{(a)}\ \begin{array}{r} 1\,0\,1 \\ 1\,1 \\ \hline \end{array} \qquad \text{(b)}\ \begin{array}{r} 1\,1\,0\,0\,1 \\ 1\,1\,1\,0 \\ \hline \end{array} \qquad \text{(c)}\ \begin{array}{r} 1\,0\,1\,1\,0\,0\,1 \\ 1\,0\,1\,0\,1\,0 \\ \hline \end{array} \qquad \text{(d)}\ \begin{array}{r} 1\,1\,0\,1\,1\,1\,0\,1\,1\,0 \\ 1\,0\,1\,1\,0\,0\,1\,1\,0\,1 \\ \hline \end{array}$$

2.1.4. The Division Algorithm

We recall that division is a process of extended subtraction. For instance, in dividing 2972 by 43, we wish to find the maximum number of times which 43 can be subtracted from 2972 and the positive number which remains after this maximum subtraction has taken place. In determining these two numbers, called **quotient** and **remainder** respectively, we proceed as follows.

We first write the two given numbers in adjacent positions horizontally with the divisor to the left of the dividend and a vertical stroke drawn to separate them. A horizontal straight line segment is drawn above the dividend thus:

$$4\,3\,\big|\overline{\,2\,9\,7\,2\,}.$$

We note that 43 is approximately 40 and that 2972 is approximately 2900. Then, since 29 is slightly more than 7 times 4, it follows that 2900 is slightly more than 70 times 40, that is, that 40 can be subtracted 70 times from 2900. Hence we may conjecture that 43 can be subtracted from 2972 *about* 70 times. We indicate this by writing 70 above the dividend with their right-hand digits vertically aligned.

$$\begin{array}{r|r} & 70 \\ \hline 43 & 2972 \end{array}$$

To verify that this is actually the case, we form the product 70 · 43 (that is, we take 43 70 times), write the result 3010 below 2972 and attempt to subtract. Finding this impossible indicates to us that our conjecture of 70 for the quotient has been an overestimate. We therefore revise our estimate downward by replacing the digit 7 by the next lower digit 6 to get 60 as our roughly estimated quotient. We now find that the product of 43 taken 60 times is 2580 and that this can be subtracted from 2972. Doing so, we find that a difference of 392 remains.

$$\begin{array}{r|r} & 60 \\ \hline 43 & 2972 \\ & 2580 \\ \cline{2-2} & 392 \end{array}$$

Since the remaining 392 is greater than the divisor 43, it is obvious that the latter can be subtracted an additional number of times. An estimate of this additional number of times may be obtained by rounding 43 off to 40 as before, and rounding 392 off to 390. Since 39 is more than 9 times 4, it follows that 390 is slightly more than 9 times 40, and hence that 392 is probably *about* 9 times 43. We record this 9 by adding it to the first quotient-estimate to obtain 69 which we now write in place of the original 60. We then verify the appropriateness of the 9 by computing 9 · 43 which we find to be 387, writing it below the remaining 392, and subtracting from the latter.

```
         69
         6̸0̸
    ________
43 | 2 9 7 2
     2 5 8 0
     -------
       3 9 2
       3 8 7
       -----
           5
```

Since this subtraction is possible we have now verified that it is possible to subtract 43 from 2972 at least 60 + 9, or 69, times. Since the difference now remaining is a number (5) which is less than 43, it is seen that it is not possible to subtract 43 an additional time, and hence 69 is seen to be the maximum number of times which 43 can be subtracted from 2972. This number 69 is then the required **quotient** and the remaining difference of 5 is the required **remainder**

If we make full use of the principle of positional notation, we may omit the writing of the zero in our quotient-estimate of 60 and also in the product (2580) of 60 times 43, and we may record the first step in the algorithm thus,

```
          6  .
    __________
43 | 2 9 7 2.
     2 5 8
     -----
       3 9 2,
```

the complete algorithmic record appearing as

```
          6 9.
    __________
43 | 2 9 7 2.
     2 5 8
     -----
       3 9 2
       3 8 7
       -----
           5.
```

While correct computation of quotient and remainder is

essential in any division problem, a correct *interpretation* of the relation of these results to the given dividend and divisor is of equal importance. The best way to ensure a proper understanding of this relation is to express it in the form of a mathematical statement of equality, or "equation", concerning the four numbers involved.

Recalling that the **remainder** (5) in the preceding problem is the number that *remains* after the **divisor** (43) has been subtracted from the **dividend** (2972) as many times as possible (namely the **quotient**, 69), the relation between these numbers may be written as

$$2972 - 69 \cdot 43 = 5$$

or, equivalently,

$$2972 = 69 \cdot 43 + 5$$

or

$$\frac{2972}{43} = \frac{69 \cdot 43 + 5}{43} = \frac{69 \cdot 43}{43} + \frac{5}{43} = 69 + \frac{5}{43}.$$

It should be emphasized that the remainder (that which *remains* after maximum subtraction has taken place) is the number 5, *not* the fraction 5/43.

If it is desired to "check", or verify, the correctness of the computed quotient and remainder, the relation

$$2972 = 69 \cdot 43 + 5$$

indicates the proper method for doing so, and the corresponding algorithm is as follows.

```
Multiply divisor →     4 3
by quotient,     →     6 9
                     -----
                     3 8 7
                   2 5 8
                   -------
                   2 9 6 7
add remainder; →         5
                   -------
                   2 9 7 2  ← compare with dividend.
```

EXERCISES

Directions for Exercises 1, 2, 3 and 4: (a) In each of these four problems obtain both quotient and remainder and identify each one clearly. (b) Also show, in the form of an equation, the relationship between dividend, divisor, quotient and remainder. (c) Check the correctness of the results in each case by multiplication of the divisor by the quotient and addition of the remainder.

1. Divide 192 by 17.
2. Divide 2245 by 47.
3. Divide 4,540,631 by 127.
4. Divide 903,010,309 by 757.
5. Describe the division algorithm without referring to any particular numerical example.
6. In the problem involving division of 2972 by 43, interpret the dividend 2972 as a number of potatoes and give two different possible interpretations of the meaning of the quotient obtained as a number of elements in a set.
7. Interpreting all numbers in this problem as octal numbers, divide 2245 by 47.
8. Interpreting all numbers in this problem as duodecimal (or base twelve) numbers, divide 2245 by 47. (Use the symbols t and e to represent the numbers which, in the decimal system, we call "ten" and "eleven" respectively; that is, use the digits 0, 1, 2, 3, 4, 5, 6, 7, 8, 9, t and e.)
9. Interpreting all numbers in this problem as binary numbers, divide 10,111,000 by 1011.
10. What is the base in the number system in which $1437 = 41 \cdot 32 + 15$?
11. What percentage of a child's time spent in learning division do you think he should be encouraged to spend in solving division problems by "trial-and-error?" Why?

2.1.5. Adaptation from Natural Numbers to Integers

Sections 2.1.1 thru 2.1.4 have been concerned with algorithms—that is, practical methods—for carrying out the operations of addition, multiplication, subtraction and division for the natural numbers.

The extension of these algorithms to the counting numbers (or non-negative integers) follows easily and immediately from our recollection of the following facts. (As a matter of fact we have already been making unconscious use of most of these facts in the preceding sections due to the inevitability of dealing with zero when describing the natural numbers by means of positional notation.)

(1) **The sum of zero and any counting number is equal to that counting number itself.** That is, **for any counting number** $\boldsymbol{n}$,

$$n + 0 = n \text{ and } 0 + n = n.$$

EXAMPLE 1

$$317 + 0 = 317, \quad 0 + 317 = 317.$$

EXAMPLE 2

$$0 + 0 = 0.$$

(2) **Any product which contains zero as a factor is equal to zero.** That is, **for any counting number** $\boldsymbol{n}$,

$$n \cdot 0 = 0 \text{ and } 0 \cdot n = 0.$$

EXAMPLE 1

$$317 \cdot 0 = 0, \quad 0 \cdot 317 = 0.$$

EXAMPLE 2

$$0 \cdot 0 = 0.$$

(3) **Zero subtracted from any counting number leaves a difference equal to that counting number itself.** That is, **for any counting number** $\boldsymbol{n}$,

$$n - 0 = n.$$

EXAMPLE 1

$$317 - 0 = 317.$$

EXAMPLE 2

$$0 - 0 = 0.$$

(4) **The difference obtained by subtracting any counting number from zero is equal to the negative of that counting number.** That is, **for any counting number** $\boldsymbol{n}$,

$$0 - n = -n.$$

EXAMPLE

$$0 - 317 = -317.$$

(5) **Zero divided by any natural number yields a quotient equal to zero.** That is, **for any natural number n,**

$$\frac{0}{n} = 0.$$

EXAMPLE

$$\frac{0}{317} = 0.$$

(6) **The operation of division is defined only for divisors which are different from zero.** That is, **for any counting number n, we do not recognize the symbol $n/0$ as having any numerical meaning.**

EXAMPLE 1

317/0 has no numerical meaning; that is, this symbol does not represent any number.

EXAMPLE 2

0/0 has no numerical meaning.

The further adaptation of the algorithms to all integers may now be accomplished by stating the relations which exist between all conceivable types of sum, product, difference and quotient which involve negative integers and those which involve the corresponding positive integers.

For addition involving negative integers we recall the following facts.

(7) **For any two positive integers m and n with m greater than n, the difference $m - n$ is a positive integer, and the sum of m and the negative of n, taken in either order, is the difference $m - n$.** That is,

$$m + (-n) = m - n \text{ and } (-n) + m = m - n.$$

EXAMPLE

$$38 + (-15) = 38 - 15 = 23$$
$$(-15) + 38 = 38 - 15 = 23.$$

(8) **For any two positive integers m and n with m less than n, the difference $m - n$ is a negative integer, namely the negative of**

the positive integer $n - m$, and the sum of m and the negative of n, taken in either order, is the negative integer $-(n - m)$. That is

$$m + (-n) = -(n - m) \text{ and } (-n) + m = -(n - m).$$

EXAMPLE

$$15 + (-38) = -(38 - 15) = -23$$
$$(-38) + 15 = -(38 - 15) = -23.$$

(9) **For any two positive integers m and n, the sum of the negative of m and the negative of n is the negative of the sum of m and n.** That is,

$$(-m) + (-n) = -(m + n).$$

EXAMPLE

$$(-34) + (-77) = -(34 + 77) = -111.$$

When many integers, some positive and some negative, are to be added, the most efficient procedure is to take advantage of the commutative and associative laws of addition to group all negative terms together first, find their sum by the repeated application of (9), and then add this sum to the sum of all the positive terms by the application of (7) or (8).

EXAMPLE

$$(-12) + (-5) + 24 + (-3) + 33 + 6 + (-1) + 4 + 2 + (-3)$$
$$= (24 + 33 + 6 + 4 + 2) + [(-12) + (-5) + (-3) + (-1) + (-3)]$$
$$= 69 + [-24] = 69 - 24 = 45.$$

For multiplication involving negative integers we make use of the following facts.

(10) **For any two positive integers m and n, the product of m and the negative of n, taken in either order, is the negative of the product of m and n.** That is,

$$m \cdot (-n) = -(m \cdot n) \text{ and } (-n) \cdot m = -(m \cdot n).$$

EXAMPLE

$$15 \cdot (-7) = -(15 \cdot 7) = -105$$
$$(-7) \cdot 15 = -(15 \cdot 7) = -105.$$

(11) **For any two positive integers *m* and *n*, the product of the negative of *m* and the negative of *n* is equal to the product of *m* and *n*.** That is,

$$(-m) \cdot (-n) = m \cdot n.$$

EXAMPLE

$$(-18) \cdot (-24) = 18 \cdot 24 = 432.$$

Any subtraction problem involving integers may be restated, by means of the following principle, in a form **involving only addition of integers.**

(12) **For any two integers *m* and *n*, the difference obtained by subtracting *n* from *m* is equal to the sum of *m* and the negative of *n*.** That is,

$$m - n = m + (-n).$$

EXAMPLE 1

$$24 - (-73) = 24 + 73 = 97.$$

EXAMPLE 2

$$(-24) - (-73) = (-24) + 73 = 73 - 24 = 49,$$

EXAMPLE 3

$$(-24) - 73 = (-24) + (-73) = -(24 + 73) = -97.$$

From principles (10) and (11) for multiplication may be deduced the following principles for division.

(13) **For any two positive integers *m* and *n*, the quotient of *m* divided by the negative of *n*, and also the quotient of the negative of *m* divided by *n*, is equal to the negative of the quotient of *m* divided by *n*.** That is,

$$\frac{m}{-n} = -\frac{m}{n} \text{ and } \frac{-m}{n} = -\frac{m}{n}.$$

EXAMPLE

$$\frac{12}{-3} = -\frac{12}{3} = -4, \frac{-12}{3} = -\frac{12}{3} = -4.$$

(14) **For any two positive integers m and n, the quotient of the negative of m divided by the negative of n is equal to the quotient of m divided by n.** That is,

$$\frac{-m}{-n} = \frac{m}{n}.$$

EXAMPLE

$$\frac{-12}{-3} = \frac{12}{3} = 4.$$

As as consequence of principles (10), (11), (13) and (14) it follows that **any extended product or product-quotient of integers is positive if it involves an even number of negative factors and negative if it involves an odd number of negative factors.** This may be seen by combining negative factors in pairs to produce positive partial products.

EXAMPLE 1

$$\begin{aligned} 5 \cdot (-4) \cdot (-3) \cdot 10 \cdot (-6) \cdot (-14) \cdot 9 \\ = [(-4) \cdot (-3)] \cdot [(-6) \cdot (-14)] \cdot 5 \cdot 10 \cdot 9 \\ = 12 \cdot 84 \cdot 5 \cdot 10 \cdot 9 = 453{,}600. \end{aligned}$$

EXAMPLE 2

$$\begin{aligned} 5 \cdot (-4) \cdot (-3) \cdot 10 \cdot (-6) \cdot (-14) \cdot 9 \cdot (-8) \\ = [(-4) \cdot (-3)] \cdot [(-6) \cdot (-14)] \cdot (-8) \cdot 5 \cdot 10 \cdot 9 \\ = 12 \cdot 84 \cdot (-8) \cdot 5 \cdot 10 \cdot 9 = 453{,}600 \cdot (-8) \\ = -3{,}628{,}800. \end{aligned}$$

EXAMPLE 3

$$\frac{5 \cdot (-3) \cdot (-14) \cdot 9 \cdot 8}{(-4) \cdot 10 \cdot (-6)} = \frac{[(-3) \cdot (-14)] \cdot [5 \cdot 9 \cdot 8]}{[(-4) \cdot (-6)] \cdot 10}$$

$$= \frac{42 \cdot 360}{24 \cdot 10} = 63.$$

EXAMPLE 4

$$\frac{5\cdot(-3)\cdot(-14)\cdot 9\cdot(-8)}{(-4)\cdot 10\cdot(-6)} = \frac{[(-3)\cdot(-14)]\cdot[5\cdot 9\cdot(-8)]}{[(-4)\cdot(-6)]\cdot 10}$$

$$= \frac{42\cdot(-360)}{24\cdot 10} = -63.$$

In particular, **a power of a negative base is positive if the exponent is an even integer, negative if the exponent is an odd integer.**

EXAMPLE 1

$$(-3)^3 = (-3)\cdot(-3)\cdot(-3) = -27.$$

EXAMPLE 2

$$(-3)^4 = (-3)\cdot(-3)\cdot(-3)\cdot(-3) = 81.$$

EXERCISES

In each of the following exercises, evaluate each expression which has numerical meaning and state which ones do not have numerical meaning.

1.

(a) $999 + 0$ (b) $0\cdot(-7)$ (c) $\frac{0\cdot 0}{0}$ (d) $\frac{0}{27}$ (e) $0 - 1$

(f) $10 - 0$ (g) $-1+0$ (h) $2\cdot 0$ (i) $0\cdot 0$ (j) $\frac{0}{2-7}$

(k) $\frac{0}{0} - \frac{0}{0}$ (l) $\frac{7}{2-2}$.

2.

(a) $(-3)+(-3)$ (b) $0-(-71)$ (c) $(-3)-(-3)$ (d) $47-(-0)$

(e) $1963 - 1776$ (f) $-20+(-4)$ (g) $78-(-61)$ (h) $-12-(-41)$

(i) $-29+13-(-61)-16+29+(-51)-0-30$

(j) $0-74+(-31)-(-13)-(-91)+(-0)+76$.

3.

(a) $24\cdot(-24)$ (b) $(-24)\cdot(-24)$ (c) $(-1)\cdot(-3)\cdot(-5)$

(d) $(-4)\cdot 0\cdot(-4)$ (e) $(-16)\cdot(-32)$ (f) $(-18)^4$ (g) $(-5)^5$ (h) 0^0

(i) $(-2)\cdot(-7)\cdot 8\cdot(-3)\cdot 5\cdot 6\cdot(-1)$

(j) $(-22)\cdot(-19)\cdot(-59)\cdot(-13)\cdot 17\cdot(-23)$.

4.

(a) $(-2292)\cdot(-343)\cdot(-4576)$ (b) $\frac{-144}{-48}$ (c) $\frac{-192}{24}$ (d) $\frac{84}{-14}$

(e) $\frac{(-92)\cdot(-14)}{-46}$ (f) $\frac{432}{(-12)\cdot(-6)}$ (g) $\frac{-576}{8\cdot 0\cdot(-36)}$

(h) $\frac{(-32) \cdot 48}{(-16) \cdot (-12)}$ (i) $\frac{4 \cdot (-6) \cdot (-14) \cdot (-12) \cdot (-18)}{7 \cdot (-8) \cdot 3 \cdot 6 \cdot (-12) \cdot (-18)}$.

(j) $\frac{(-91) \cdot 12 \cdot (-35) \cdot 42 \cdot 39 \cdot (-14)}{(-9) \cdot (-15) \cdot (-4) \cdot (-13) \cdot (-7)}$.

5.
(a) $5 - 2 \cdot 3$ (b) $(5 - 2) \cdot 3$ (c) $5 \cdot (-2) \cdot 3$ (d) $18 - 6/3$
(e) $(18 - 6)/3$ (f) $18 \cdot (-6/3)$ (g) $[18 \cdot (-6)]/3$ (h) $6 - 18/3$
(i) $\frac{4 + 2 \cdot (-2)}{(-2) - (-2)}$ (j) $\frac{(-3) + 3}{5 - (-5)}$.

2.1.6. Algorithms for Greatest Common Divisor

Since $84 = 2 \cdot 42$ and hence $84/2$ is the integer 42, the integer 2 is called a **divisor** of the integer 84. Similarly, since $48 = 2 \cdot 24$ and hence $48/2$ is the integer 24, 2 is also a divisor of 48. Since 2 is a divisor of both 84 and 48 it is called a **common divisor** of 84 and 48. It may be noted that 4, 6 and 12 are also common divisors of 84 and 48 but that there are no other integers which are common divisors of 84 and 48. Since 12 is the largest of the set of common divisors of 84 and 48 it is called the **greatest common divisor** (or **g.c.d.**) of 84 and 48. [The expression **highest common factor** (or **h.c.f.**) is sometimes used as an equivalent for "greatest common divisor".]

In simple cases the g.c.d. of two or more numbers may be found by expressing the numbers in factored form and comparing their factors.

EXAMPLE 1

$$84 = 2^2 \cdot 3 \cdot 7$$
$$48 = 2^4 \cdot 3.$$

It is seen directly that $2^2 \cdot 3$, or 12, is a divisor of both 84 and 48 and also that 12 is the largest possible divisor of these two numbers; hence the g.c.d. of 84 and 48 is 12.

EXAMPLE 2

To find the g.c.d. of the three numbers 120, 384 and 576 we observe their factored forms,

$$120 = 2^3 \cdot 3 \cdot 5$$
$$384 = 2^7 \cdot 3$$
$$576 = 2^6 \cdot 3^2,$$

and note that $2^3 \cdot 3$, or 24, is a divisor which is common to each of the given numbers and is the largest such common divisor; thus the g.c.d. of 120, 384 and 576 is 24.

For determining g.c.d.s in more complicated cases, a method discovered and used by Euclid over 2000 years ago is most convenient and useful. It is illustrated by the following example using the numbers 84 and 48.

First we divide the larger of the two numbers by the smaller to obtain a quotient and a remainder, thus:

```
       1 ← Quotient
    ____
48 | 84
     48
     __
     36 ← Remainder.
```

Next we divide the previous divisor, 48, by the remainder, 36, to obtain a new quotient and a new remainder, thus:

```
       1 ← Quotient
    ____
36 | 48
     36
     __
     12 ← Remainder.
```

We continue in this manner, obtaining finally a zero remainder.

```
       3 ← Quotient
    ____
12 | 36
     36
     __
      0 ← Remainder.
```

It may be observed that the last remainder which is different from zero, namely 12, is the g.c.d. of 84 and 48. It can be proved that this is no coincidence but that the last non-zero remainder obtained in a sequential procedure of this sort must always be the g.c.d. of the two numbers used in the first division.

This Euclidean algorithm, as it is called, is most valuable when dealing with large numbers whose factors may all be large, unknown and not easy to find. No matter how large the factors may be, the Euclidean algorithm provides a straightforward method which may be carried out in a routine manner and does not depend on any trial-and-error procedure of any sort.

The ability to find the g.c.d. of two numbers is useful in connection with the reduction of a fraction to lowest terms and in other applications. Following is an example of its use in reducing a fraction to lowest terms.

Suppose we wish to reduce the fraction 111,384/29,700 to lowest terms. Since this can be done by dividing both numerator and denominator by their g.c.d. it is worthwhile to find this number. We do so.

$$\begin{array}{r|r} & 3 \\ \hline 29700 & 111384 \\ & 89100 \\ \hline & 22284 \end{array}$$

$$\begin{array}{r|r} & 1 \\ \hline 22284 & 29700 \\ & 22284 \\ \hline & 7416 \end{array}$$

$$\begin{array}{r|r} & 3 \\ \hline 7416 & 22284 \\ & 22248 \\ \hline & 36 \end{array}$$

$$
\begin{array}{r|l}
 & 206 \\ \hline
36 & 7416 \\
 & 72 \\ \cline{2-2}
 & 216 \\
 & 216 \\ \cline{2-2}
 & 0.
\end{array}
$$

Hence the g.c.d. of 11,384 and 29,700 is 36 and

$$\frac{111,384}{29,700} = \frac{111,384/36}{29,700/36} = \frac{3094}{825}.$$

Sometimes it is necessary to find the g.c.d. of several numbers. This may be done by finding the g.c.d. of some two of them, then finding the g.c.d. of a third number and the previously obtained g.c.d. of the first two, and so on. This procedure is illustrated in the following example involving the determination of the g.c.d. of 111,384, 29,700, 1728 and 3762.

From the preceding example we know that the g.c.d. of 11,384 and 29,700 is 36. We now find the g.c.d. of 36 and 1728.

$$
\begin{array}{r|l}
 & 48 \\ \hline
36 & 1728 \\
 & 144 \\ \cline{2-2}
 & 288 \\
 & 288 \\ \cline{2-2}
 & 0.
\end{array}
$$

Since this shows that 36 itself is a divisor of 1728, we now know that the g.c.d. of 11,384, 29,700 and 1728 is 36. We next find the g.c.d. of 36 and 3762.

$$
\begin{array}{r|l}
 & 104 \\ \hline
36 & 3762 \\
 & 36 \\ \cline{2-2}
 & 162 \\
 & 144 \\ \cline{2-2}
 & 18
\end{array}
$$

$$
\begin{array}{r|l}
 & 2 \\ \hline
18 & 36 \\
 & 36 \\ \cline{2-2}
 & 0.
\end{array}
$$

Thus 18 is the g.c.d. of 36 and 3762 and hence it is also the g.c.d of 11,384, 29,700, 1728 and 3762.

EXERCISES

1. Describe the Euclidean g.c.d. algorithm without referring to any particular numerical example.
2. Give a formal description (without referring to any particular example) of the factor method of finding the g.c.d. of several integers.

In each of the following problems use both the Euclidean algorithm and the method of factors.

3. Find the g.c.d. of: (a) 384 and 648; (b) 275 and 3640.
4. Find the g.c.d. of: (a) 899 and 191; (b) 29,625 and 842,375.
5. Find the g.c.d. of 3255, 1470, 1680 and 21,000.
6. Find the g.c.d. of 242,864, 872,924 and 1,864,984.

2.1.7. Algorithm for Least Common Multiple

A **multiple** of an integer n is any product of integers in which n is one of the factors. As examples, 6, 12, 18, 24, 30, 36, 42, 48, 54, etc., are all multiples of 6. Similarly, 8, 16, 24, 32, 40, 48, 56, etc., are all multiples of 8. We notice that among the numbers listed, some, namely 24 and 48, are multiples of both 6 and 8. These numbers (and others such as 72, 96, 120, etc.) are called

common multiples of 6 and 8. Among all such common multiples of 6 and 8 the number 24 is noteworthy as the smallest, or least, and it is therefore called the **least common multiple** of 6 and 8. (The expression "least common multiple" is frequently abbreviated as l.c.m.)

A knowledge of least common multiples (and the ability to find them readily) becomes important in connection with the addition of rational numbers.

For an integer to be a multiple of 6, it must be a product of the form $6 \cdot n$, or $2 \cdot 3 \cdot n$, where n is also an integer. Hence any such multiple must contain 6 as a factor, and hence must also contain as factors the factors 2 and 3 of 6. Similarly, for an integer to be a multiple of 8, it must be a product of the form $8 \cdot m$, or $2^3 \cdot m$, for some integer m. Thus it must contain 8 as a factor; that is, it must contain the factor 2 three times. We can now deduce that for a number to be a common multiple of 6 and 8 it must contain the factors 2 and 3 so that it can be expressed as $2 \cdot 3 \cdot n$, or $6 \cdot n$, and it also must contain the factor 2 three times so that it can be expressed as $2^3 \cdot m$, 8 or $\cdot\, m$. Thus any common multiple of 6 and 8 must contain at least the factors 2^3 and 3; finally, if it is to be the *least* (or smallest) common multiple of 6 and 8, it should contain no other factor than these minimally necessary ones. Thus the least common multiple of 6 and 8 is determined as $2^3 \cdot 3$, or 24.

The l.c.m. algorithm may be summarized as follows. **To find the l.c.m. of a given set of integers, first express each of the given integers as a product of its prime factors, then form a product which contains each prime factor which appears in any of the given integers and such that each such factor occurs in the constructed product only as many times as the greatest number of times which it occurs in any one of the given integers.**

Example

To find the l.c.m. of 84 and 144, we first observe that

$$84 = 2^2 \cdot 3 \cdot 7 \text{ and } 144 = 2^4 \cdot 3^2.$$

The factors 2, 3 and 7 occur in these factored forms of the given integers, so each of them will occur also in the required l.c.m. Since the greatest number of times that 2 occurs in any one of the given integers is four, we therefore include 2^4 in the l.c.m. Similarly, since the greatest number of times that 3 occurs in any one of the given integers is two, we include 3^2 in the l.c.m. Finally, since 7 occurs only once, we include 7^1, or 7, in the l.c.m. Thus the required l.c.m. is $2^4 \cdot 3^2 \cdot 7$ or 1008.

The summarized algorithm stated above applies equally well for finding the l.c.m. of any number of given integers.

EXAMPLE

To find the l.c.m. of 72, 162, 546 and 728 we first observe that

$$72 = 2^3 \cdot 3^2,\ 162 = 2 \cdot 3^4,\ 546 = 2 \cdot 3 \cdot 7 \cdot 13, \text{ and } 728 = 2^3 \cdot 7 \cdot 13.$$

The required l.c.m. is then seen to be $2^3 \cdot 3^4 \cdot 7 \cdot 13$ or 58,968.

EXERCISES

1. Find the l.c.m. of: (a) 56 and 96; (b) 432 and 192.

2. Find the l.c.m. of: (a) 1728 and 1296; (b) 20,375 and 3125.

3. Find the l.c.m. of: (a) 18, 42 and 72; (b) 75, 130 and 550.

4. Find the l.c.m. of 144, 30, 96, 42 and 105.

5. Select any two positive integers. (To make the problem more interesting, choose numbers between 100 and 1000.) (a) Find the g.c.d. of the chosen integers. (b) Find the l.c.m. of the chosen integers. (c) Compute the product of the g.c.d. and the l.c.m. of the chosen integers. (d) Compute the product of the chosen integers themselves. (e) Compare the results in (c) and (d). (f) From your observations in (e), what conjecture do you feel inspired to make concerning the product of any two positive integers and the product of their g.c.d. and l.c.m.?

6. Suggest an efficient algorithm for finding the l.c.m. of two positive integers by making use of their g.c.d.

2.1.8. Algorithms for Change of Base

There are two important considerations involved in change-of-base problems. One is that most such problems are concerned either with changing from the base-ten notation to some other one, or vice versa; changes between two bases both different from ten are relatively less frequent. The second consideration is that, while the actual arithmetical computation involved in changing from one notation to another could be carried on in any system, it is nevertheless a cogent fact that most of us are accustomed to doing arithmetical work rapidly and accurately only in the base-ten notation; hence it is desirable that we use this familiar system for the arithmetic of our change-of-base computations, even tho we might be changing from base eight to base nineteen.

First let's consider the problem of changing a number given in base-ten notation, say 28,394, to its corresponding representation in base-seven notation. This is the problem of finding a set of digits d_0, d_1, d_2, d_3, etc. (whose individual values may range from 0 to 6 inclusive) such that

$$28{,}394 = d_0 + d_1 \cdot 7 + d_2 \cdot 7^2 + d_3 \cdot 7^3 + \dots .$$

We notice that, in the base seven expression at the right, each term after the first one is divisible by 7; hence if the entire expression is divided by 7, the first term, d_0, will be the remainder and the quotient will be

$$d_1 + d_2 \cdot 7 + d_3 \cdot 7^2 + \dots .$$

Similarly, if this quotient be divided by 7, the remainder will be d_1 and the quotient will be

$$d_2 + d_3 \cdot 7 + \dots .$$

If this quotient in turn be divided by 7, the remainder will be d_2 and the new quotient will be

$$d_3 + \dots ,$$

and so on. But since the expression we started with is a representation for 28,394, all the statements we have made about remainders and quotients after dividing by 7 apply to the number 28,394 and its successive quotients. Thus the required digits

d_0, d_1, d_2, d_3, etc., are obtained as the remainders in the repeated division by 7 of 28,394 and its successive quotients thus:

$$\begin{array}{r@{\,}l} & 4056 \\ 7 & \overline{)28394} \\ & 28 \\ \hline & 039 \\ & 35 \\ \hline & 44 \\ & 42 \\ \hline & 2 = d_0 \end{array} \qquad \begin{array}{r@{\,}l} & 579 \\ 7 & \overline{)4056} \\ & 35 \\ \hline & 55 \\ & 49 \\ \hline & 66 \\ & 63 \\ \hline & 3 = d_1 \end{array}$$

$$\begin{array}{r@{\,}l} & 82 \\ 7 & \overline{)579} \\ & 56 \\ \hline & 19 \\ & 14 \\ \hline & 5 = d_2 \end{array} \qquad \begin{array}{r@{\,}l} & 11 \\ 7 & \overline{)82} \\ & 7 \\ \hline & 12 \\ & 7 \\ \hline & 5 = d_3 \end{array}$$

$$\begin{array}{r@{\,}l} & 1 \\ 7 & \overline{)11} \\ & 7 \\ \hline & 4 = d_4 \end{array} \qquad \begin{array}{r@{\,}l} & 0 \\ 7 & \overline{)1} \\ & 0 \\ \hline & 1 = d_5. \end{array}$$

Thus the base seven notation for the base ten number 28,394 is

$$\begin{aligned} d_0 + d_1 \cdot 7 + d_2 \cdot 7^2 + d_3 \cdot 7^3 + d_4 \cdot 7^4 + d_5 \cdot 7^5 \\ &= d_5 \cdot 7^5 + d_4 \cdot 7^4 + d_3 \cdot 7^3 + d_2 \cdot 7^2 + d_1 \cdot 7 + d_0 \\ &= 1 \cdot 7^5 + 4 \cdot 7^4 + 5 \cdot 7^3 + 5 \cdot 7^2 + 3 \cdot 7 + 2 \\ &= 145{,}532. \end{aligned}$$

To make clearer the relation of the original base-ten representation of the number involved and its new base-seven representation it is customary to attach a subscript to each representation to indicate the base being used. Thus the above result might be written as

$$28{,}394_{10} = 145{,}532_7$$

or, if there were any danger of misinterpreting the meaning of the subscript "$_{10}$", as

$$28{,}394_{\text{ten}} = 145{,}532_{\text{seven}}.$$

Since all divisions performed in the determination of the base-seven digits for this problem (and similar problems) are simple ones we could write them as follows.

```
7 | 2 8 3 9 4
   ----------
    4 0 5 6;  2   (where the 2 at the right of the semi-colon
 7 | 4 0 5 6          represents the remainder)
   ---------
      5 7 9;  3
   7 | 5 7 9
     -------
        8 2;  5
     7 | 8 2
       -----
        1 1;  5
     7 | 1 1
       -----
          1
          1;  4
       7 | 1
         ---
          0;  1.
```

Examination of this somewhat condensed computational record shows that each quotient has been written twice. This is extravagant and unnecessary; there is no reason why we cannot shorten the record to the following form.

$$\begin{array}{rrl}
7 \,|\, & 2\,8\,3\,9\,4 & \\
7 \,|\, & 4\,0\,5\,6; & 2 \\
7 \,|\, & 5\,7\,9; & 3 \\
7 \,|\, & 8\,2; & 5 \\
7 \,|\, & 1\,1; & 5 \\
7 \,|\, & 1; & 4 \\
 & 0; & 1.
\end{array}$$

Since our problem is one of changing to base-seven notation, we know that each divisor is 7; hence there is no need to clutter the computational record with repeated mention of 7. Neither is there any real need for the lines which separate the various numbers; nor for the semi-colons. The essential information is all given by simply writing down the successive quotients and remainders thus:

$$\begin{array}{rl}
2\,8\,3\,9\,4 & \\
4\,0\,5\,6 & 2 \\
5\,7\,9 & 3 \\
8\,2 & 5 \\
1\,1 & 5 \\
1 & 4 \\
0 & 1.
\end{array}$$

Noting that the final quotient is necessarily zero and the final remainder is equal to the last non-zero quotient we see that there is no necessity for writing down the zero quotient and the remainder in the last line. We can, therefore, condense the algorithmic record to the form

$$\begin{array}{rll}
2\,8\,3\,9\,4 & & \\
4\,0\,5\,6 & 2 & \leftarrow \text{First remainder} \\
5\,7\,9 & 3 & \leftarrow \text{Second remainder} \\
8\,2 & 5 & \leftarrow \text{Third remainder} \\
1\,1 & 5 & \leftarrow \text{Fourth remainder} \\
1 & 4 & \leftarrow \text{Next-to-last remainder} \\
\nwarrow & & \text{—— Last remainder.}
\end{array}$$

The remainders, read in the order opposite to that in which they were obtained, form the digits in the base-seven representation 145,532 for the given base-ten number 28,394.

EXAMPLE

Change of $97{,}426_{10}$ to a base-four number

$$\begin{array}{rr} 97426 & \\ 24356 & 2 \\ 6089 & 0 \\ 1522 & 1 \\ 380 & 2 \\ 95 & 0 \\ 23 & 3 \\ 5 & 3 \\ 1 & 1. \end{array}$$

Then $97{,}426_{10} = 113{,}302{,}102_4$.

The key to the inverse problem of finding the base-ten representation of the base-seven number 145,532 is found by noting that the latter number

$$1 \cdot 7^5 + 4 \cdot 7^4 + 5 \cdot 7^3 + 5 \cdot 7^2 + 3 \cdot 7 + 2$$

can be expressed as

$$((((1 \cdot 7 + 4) \cdot 7 + 5) \cdot 7 + 5) \cdot 7 + 3) \cdot 7 + 2$$

and can be obtained by the successive computation of the quantities

$$\begin{array}{l} 1 \\ 1 \cdot 7 \\ 1 \cdot 7 + 4 \\ (1 \cdot 7 + 4) \cdot 7 \\ (1 \cdot 7 + 4) \cdot 7 + 5 \\ ((1 \cdot 7 + 4) \cdot 7 + 5) \cdot 7 \\ ((1 \cdot 7 + 4) \cdot 7 + 5) \cdot 7 + 5 \\ (((1 \cdot 7 + 4) \cdot 7 + 5) \cdot 7 + 5) \cdot 7 \\ (((1 \cdot 7 + 4) \cdot 7 + 5) \cdot 7 + 5) \cdot 7 + 3 \\ ((((1 \cdot 7 + 4) \cdot 7 + 5) \cdot 7 + 5) \cdot 7 + 3) \cdot 7 \\ ((((1 \cdot 7 + 4) \cdot 7 + 5) \cdot 7 + 5) \cdot 7 + 3) \cdot 7 + 2. \end{array}$$

We might, at first, do this as follows.

$$\begin{array}{r@{\;}l}
1 & = 1 \\[6pt]
7 & \\
\hline
7 & = 1 \cdot 7 \\[6pt]
4 & \\
\hline
1\,1 & = 1 \cdot 7 + 4 \\[6pt]
7 & \\
\hline
7\,7 & = (1 \cdot 7 + 4) \cdot 7 \\[6pt]
5 & \\
\hline
8\,2 & = (1 \cdot 7 + 4) \cdot 7 + 5 \\[6pt]
7 & \\
\hline
5\,7\,4 & = ((1 \cdot 7 + 4) \cdot 7 + 5) \cdot 7 \\[6pt]
5 & \\
\hline
5\,7\,9 & = ((1 \cdot 7 + 4) \cdot 7 + 5) \cdot 7 + 5 \\[6pt]
7 & \\
\hline
4\,0\,5\,3 & = (((1 \cdot 7 + 4) \cdot 7 + 5) \cdot 7 + 5) \cdot 7 \\[6pt]
3 & \\
\hline
4\,0\,5\,6 & = (((1 \cdot 7 + 4) \cdot 7 + 5) \cdot 7 + 5) \cdot 7 + 3 \\[6pt]
7 & \\
\hline
2\,8\,3\,9\,2 & = ((((1 \cdot 7 + 4) \cdot 7 + 5) \cdot 7 + 5) \cdot 7 + 3) \cdot 7 \\[6pt]
2 & \\
\hline
2\,8\,3\,9\,4 & = ((((1 \cdot 7 + 4) \cdot 7 + 5) \cdot 7 + 5) \cdot 7 + 3) \cdot 7 + 2.
\end{array}$$

This could be shortened considerably by leaving out the multiplier 7 altogether. Writing the number 145,532 once for reference, the computational record could be shortened to the following form.

$$
\begin{array}{ll}
1\,4\,5\,5\,3\,2 & \\
\;\;7 & = 1 \cdot 7 \\
\hline
1\,1 & = 1 \cdot 7 + 4 \\
\;\;7\,7 & = (1 \cdot 7 + 4) \cdot 7 \\
\hline
\;\;8\,2 & = (1 \cdot 7 + 4) \cdot 7 + 5 \\
\;\;5\,7\,4 & = ((1 \cdot 7 + 4) \cdot 7 + 5) \cdot 7 \\
\hline
\;\;5\,7\,9 & = ((1 \cdot 7 + 4) \cdot 7 + 5) \cdot 7 + 5 \\
\;\;4\,0\,5\,3 & = (((1 \cdot 7 + 4) \cdot 7 + 5) \cdot 7 + 5) \cdot 7 \\
\hline
\;\;4\,0\,5\,6 & = (((1 \cdot 7 + 4) \cdot 7 + 5) \cdot 7 + 5) \cdot 7 + 3 \\
\;\;2\,8\,3\,9\,2 & = ((((1 \cdot 7 + 4) \cdot 7 + 5) \cdot 7 + 5) \cdot 7 + 3) \cdot 7 \\
\hline
\;\;2\,8\,3\,9\,4 & = ((((1 \cdot 7 + 4) \cdot 7 + 5) \cdot 7 + 5) \cdot 7 + 3) \cdot 7 + 2.
\end{array}
$$

If the addition of the digits 4, 5, 5, 3 and 2 be done mentally following each successive multiplication by 7 the above record can be further shortened to the following form.

$$
\begin{array}{l}
1\,4\,5\,5\,3\,2 \\
1\,1 \\
\;\;8\,2 \\
\;\;5\,7\,9 \\
\;\;4\,0\,5\,6 \\
\;\;2\,8\,3\,9\,4.
\end{array}
$$

EXAMPLE

Change of $1{,}233{,}212_4$ to a base-ten number

$$
\begin{array}{l}
1\,2\,3\,3\,2\,1\,2 \\
\;\;6 \\
\;\;2\,7 \\
\;\;1\,1\,1 \\
\;\;\;\;4\,4\,6 \\
\;\;\;\;1\,7\,8\,5 \\
\;\;\;\;\;\;7\,1\,4\,2.
\end{array}
$$

Hence $1{,}233{,}212_4 = 7{,}142_{10}$.

EXERCISES

1. Change 396_{10} to the corresponding representation in the positional notation system with base: (a) 8; (b) 5; (c) 3.

2. Change 4197_{10} to the corresponding representation in the positional notation system with base: (a) 9; (b) 6; (c) 2.

3. Change $32{,}942_{10}$ to the corresponding representation in the positional notation system with base twelve. Use *t* for ten and *e* for eleven in this system.

4. For a positive integer *n* different from either one or ten explain the algorithm for changing the base-ten representation of any positive integer to its base-*n* representation. Do this without referring to any particular example.

5. Obtain the base-ten representations for the following numbers:
(a) $10{,}111{,}010_2$; (b) $1{,}110{,}101{,}100_2$; (c) $1{,}010{,}111{,}110_2$.

6. Obtain the base-ten representations for the following numbers:
(a) $12{,}102{,}120_3$; (b) $212{,}201_3$; (c) $10{,}102{,}022_3$.

7. Obtain the base-ten representations for the following numbers:
(a) $312{,}032_4$; (b) $1{,}230{,}321_4$; (c) $33{,}333_4$.

8. Obtain the base-ten representations for the following numbers:
(a) $41{,}203_5$; (b) $4{,}242{,}310_5$: (c) $121{,}034_5$.

9. Obtain the base-ten representations for the following numbers:
(a) $42{,}627_8$; (b) $232{,}427_8$; (c) $76{,}5432{,}10_8$.

10. Obtain the base-ten representations for the following numbers:
(a) $82{,}753_9$; (b) $276{,}480_9$; (c) $37{,}085{,}241_9$.

11. Obtain the base-ten representation for $2e{,}610{,}t93{,}5e4_{12}$, where *t* and *e* are the digit symbols for ten and eleven respectively.

12. For a positive integer *n* different from either one or ten explain the algorithm for changing the base-*n* representation of any positive integer to its base-ten representation. Do this without referring to any particular example.

13. Change $452{,}031_7$ to a base-six representation.

14. Make a table showing the base-*n* representation of the number *m* with *m* taking each integral value from one to nine inclusive and *n* taking each integral value from two to nine inclusive.

CHAPTER 2.2

COMPUTATION WITH RATIONAL NUMBERS

THERE is so little that is new in the computational algorithms for the rational numbers that the topic scarcely deserves the separate mention which we have given it. Actually—and it is fortunate for us that this is so—all these algorithms are direct applications of the structural principles explained in section 1.4.3.3.1 and of the computational algorithms for integers.

2.2.1. Multiplication

We have shown in section 1.4.3.3.1 that **the product of two rational numbers is that rational number whose numerator is the product of their numerators and whose denominator is the product of their denominators.** Stated more briefly in symbolic form, **for any two rational numbers a/b and c/d,**

$$\frac{a}{b}\cdot\frac{c}{d}=\frac{a\cdot c}{b\cdot d}$$

EXAMPLE

$$\frac{24}{143}\cdot\frac{55}{82}=\frac{24\cdot 55}{143\cdot 82}\,.$$

It should be understood, of course, that the product would generally not be left in the form just given but would be reduced to lowest terms by determining the greatest common divisor of numerator and denominator and then removing it, in accordance with the principle of pages 203 and 205, thus:

$$\frac{24}{143}\cdot\frac{55}{82}=\frac{24\cdot 55}{143\cdot 82}=\frac{2^3\cdot 3\cdot 5\cdot 11}{2\cdot 11\cdot 13\cdot 41}$$
$$=\frac{(2\cdot 11)\cdot(2^2\cdot 3\cdot 5)}{(2\cdot 11)\cdot(13\cdot 41)}=\frac{2^2\cdot 3\cdot 5}{13\cdot 41}=\frac{60}{533}.$$

2.2.2. Addition

From section 1.4.3.3.1 we know that **the sum of two rational numbers with a common denominator** (that is, having the same denominator) **is that rational number whose numerator is the sum of their numerators and whose denominator is their common denominator.** In symbols: **for any two rational numbers a/b and c/b,**

$$\frac{a}{b}+\frac{c}{b}=\frac{a+c}{b}.$$

EXAMPLE

$$\frac{16}{21}+\frac{2}{21}=\frac{16+2}{21}=\frac{18}{21}=\frac{2\cdot 3^2}{3\cdot 7}=\frac{2\cdot 3}{7}=\frac{6}{7}.$$

The sum of two rational numbers with different denominators is obtained by applying the preceding algorithm after first converting each of the given rational numbers into equivalent forms having as a common denominator the least common multiple of the original denominators. [This l.c.m. is called the **least common denominator** (or l.c.d.) for the given rational numbers.]

EXAMPLE 1

$$\frac{31}{42}+\frac{13}{105}=\frac{31}{2\cdot 3\cdot 7}+\frac{13}{3\cdot 5\cdot 7}$$
$$=\frac{31\cdot 5}{(2\cdot 3\cdot 7)\cdot 5}+\frac{13\cdot 2}{(3\cdot 5\cdot 7)\cdot 2}$$
$$=\frac{155}{2\cdot 3\cdot 5\cdot 7}+\frac{26}{2\cdot 3\cdot 5\cdot 7}=\frac{155+26}{2\cdot 3\cdot 5\cdot 7}$$
$$=\frac{181}{2\cdot 3\cdot 5\cdot 7}=\frac{181}{210}.$$

EXAMPLE 2

$$\frac{-31}{42}+\frac{13}{105}=\frac{-31}{2\cdot 3\cdot 7}+\frac{13}{3\cdot 5\cdot 7}$$

$$=\frac{-31\cdot 5}{(2\cdot 3\cdot 7)\cdot 5}+\frac{13\cdot 2}{(3\cdot 5\cdot 7)\cdot 2}$$

$$=\frac{-155}{2\cdot 3\cdot 5\cdot 7}+\frac{26}{2\cdot 3\cdot 5\cdot 7}=\frac{-155+26}{2\cdot 3\cdot 5\cdot 7}$$

$$=\frac{-129}{2\cdot 3\cdot 5\cdot 7}$$

$$=\frac{-(3\cdot 43)}{2\cdot 3\cdot 5\cdot 7}=\frac{3\cdot(-43)}{3\cdot 2\cdot 5\cdot 7}=\frac{-43}{2\cdot 5\cdot 7}$$

$$=\frac{-43}{70}=-\frac{43}{70}.$$

2.2.3. Subtraction

In view of the fact that the subtraction of any rational number can be shown to have the same effect as the addition of its negative, the algorithm stated in section 2.2.2 for the addition of rational numbers effectively covers the case for the subtraction of rational numbers also. However, if a formal statement of the algorithm for subtraction of rational numbers is desired, it may be given as follows.

The difference of two rational numbers with a common denominator is that rational number whose numerator is the difference of their numerators and whose denominator is their common denominator. In symbols: **for any two rational numbers a/b and c/b,**

$$\frac{a}{b}-\frac{c}{b}=\frac{a-c}{b}$$

EXAMPLE

$$\frac{375}{288} - \frac{129}{288} = \frac{375 - 129}{288} = \frac{246}{288} = \frac{2 \cdot 3 \cdot 41}{2^5 \cdot 3^2}$$

$$= \frac{(2 \cdot 3) \cdot 41}{(2 \cdot 3) \cdot 2^4 \cdot 3} = \frac{41}{2^4 \cdot 3} = \frac{41}{48}.$$

The difference of two rational numbers with different denominators is obtained by applying the preceding algorithm after first converting each of the given rational numbers into equivalent forms having as a common denominator the least common multiple of the original denominators.

EXAMPLE

$$\frac{4}{33} - \frac{19}{21} = \frac{4}{3 \cdot 11} - \frac{19}{3 \cdot 7} = \frac{4 \cdot 7}{3 \cdot 7 \cdot 11} - \frac{19 \cdot 11}{3 \cdot 7 \cdot 11}$$

$$= \frac{4 \cdot 7 - 19 \cdot 11}{3 \cdot 7 \cdot 11}$$

$$= \frac{28 - 209}{3 \cdot 7 \cdot 11} = \frac{-181}{3 \cdot 7 \cdot 11} = -\frac{181}{231}.$$

2.2.4. Division

The quotient of two rational numbers is that rational number whose numerator is the product of the numerator of the dividend and the denominator of the divisor and whose denominator is the product of the denominator of the dividend and the numerator of the divisor. Or, in symbols: **for any two rational numbers a/b and c/d,**

$$\frac{a/b}{c/d} = \frac{a \cdot d}{b \cdot c}.$$

EXAMPLE

$$\frac{27/85}{42/39} = \frac{27}{85} \cdot \frac{39}{42} = \frac{3^3 \cdot (3 \cdot 13)}{(5 \cdot 17) \cdot (2 \cdot 3 \cdot 7)}$$

$$= \frac{3 \cdot (3^3 \cdot 13)}{3 \cdot (2 \cdot 5 \cdot 7 \cdot 17)} = \frac{3^3 \cdot 13}{2 \cdot 5 \cdot 7 \cdot 17} = \frac{351}{1190}.$$

EXERCISES

1. Compute the following products. Remember that reduction of the result to the simplest possible form is part of the job.

(a) $\frac{84}{48} \cdot \frac{4}{7}$ (b) $\frac{108}{39} \cdot \frac{65}{48}$ (c) $\frac{292}{27} \cdot \frac{14}{57}$ (d) $\frac{12}{135} \cdot \frac{112}{288}$

(e) $\frac{432}{576} \cdot \frac{195}{224}$.

2. Compute the following sums. Remember to reduce results to lowest terms.

(a) $\frac{4}{9} + \frac{9}{4}$ (b) $\frac{5}{32} + \frac{7}{16}$ (c) $\frac{13}{24} + \frac{37}{60}$ (d) $\frac{214}{192} + \frac{41}{144}$

(e) $\frac{231}{1728} + \frac{168}{435}$.

3. Compute the following differences. Reduce results to lowest terms.

(a) $\frac{17}{28} - \frac{9}{70}$ (b) $\frac{121}{42} - \frac{84}{120}$ (c) $\frac{35}{72} - \frac{59}{48}$ (d) $\frac{82}{110} - \frac{47}{63}$

(e) $\frac{118}{228} - \frac{452}{832}$.

4. Compute the following quotients. Reduce results to lowest terms.

(a) $\frac{2/3}{3/2}$ (b) $\frac{14/15}{95/63}$ (c) $\frac{192/125}{15/288}$ (d) $\frac{1/72}{1/96}$ (e) $\frac{858/324}{186/912}$.

5. Perform the indicated computations. Express results in simplest possible form.

(a) $\frac{26}{42} + \frac{192}{56} - \frac{108}{98}$ (b) $(\frac{18}{28} - \frac{69}{56}) \cdot (\frac{42}{64} + \frac{39}{72})$

(c) $\frac{432}{297} \cdot \frac{22}{112} - \frac{63}{128} \cdot \frac{144}{81}$ (d) $\frac{(224/87) - (195/168)}{(32/91) - (111/68)}$

(e) $\frac{225/64}{49/72} + \frac{84}{111}$.

6. (a) Express the sum $a/b + c/d$ as a single fraction. (b) Express the difference $a/b - c/d$ as a single fraction. (c) Express the sum $a/b + c/d + e/f$ as a single fraction in simplest form. (d) Express the sum $a/b + c/d + e/f + g/h$ as a single fraction in simplest form.

CHAPTER 2.3

COMPUTATION WITH REAL NUMBERS; DECIMAL REPRESENTATION

EVERY real number is either a rational number or an irrational number. The addition, multiplication, subtraction and division of the rational numbers can be performed, of course, by the algorithms of Chapter 2.2. While these methods cannot be applied directly to computation with irrational numbers, we shall nevertheless show that the rational number methods of computation can be adapted to the solution of all computational problems with irrational numbers in a manner that is satisfactory for practical purposes.

The basic principle that makes it possible for us to solve problems involving computation with irrational numbers by using the computational algorithms for rational numbers is that **for any irrational number there can be found a rational number which differs from it by less than an amount as small as we may wish to specify**; such a rational number is therefore said to be "approximately equal" to the irrational number. For instance, the "square root of two" ($2^{1/2}$) is an irrational number but from the comparisons shown on page 121 it is seen that the rational number 1.414 differs from $2^{1/2}$ by less than .001, the rational number 1.414213 differs from $2^{1/2}$ by less than .000001, the rational number 1.414213562 differs from $2^{1/2}$ by less than .000000001, and so on. **Computations with irrational numbers can then be replaced by the same computations with approximately equal rational numbers to obtain sums, products, powers, differences,**

quotients, etc., which are approximately equal to the corresponding results for the given irrational numbers.

The form in which rational approximations to irrational numbers are most often written is the decimal notation discussed in section 1.4.4.6. As a matter of fact, rational numbers in terminating decimal form are used not only as approximations to irrational numbers but also as convenient approximations to rational numbers whose decimal forms are of the non-terminating repeating type.

2.3.1. Addition

The practical method of (approximate) addition of real numbers by the method of decimal approximation is illustrated by the following problem.

EXAMPLE

Computation of the sums $(25/8) + (34/3) + 2^{1/2} + \pi$ accurately to four decimal places.

As a reasonable insurance of obtaining the result with the required accuracy we shall use *five* decimal places during the computation and then "round off" the final result to four decimal places. Since

$$\frac{25}{8} = 3.12500 = 312{,}500. \cdot 10^{-5}$$

$$\frac{34}{3} \approx\dagger\ 11.33333 = 1{,}133{,}333. \cdot 10^{-5}$$

$$2^{1/2} \approx 1.41421 = 141{,}421. \cdot 10^{-5}$$

$$\pi \approx 3.14159 = 314{,}159. \cdot 10^{-5},$$

then by the distributive law we have

$$\frac{25}{8} + \frac{34}{3} + 2^{1/2} + \pi \approx 312{,}500. \cdot 10^{-5} + 1{,}133{,}333. \cdot 10^{-5} + 141{,}421. \cdot 10^{-5} + 314{,}159. \cdot 10^{-5}$$

† The symbol "$\approx$" means "is approximately equal to".

$$= (312{,}500. + 1{,}133{,}333. + 141{,}421. + 314{,}159.) \cdot 10^{-5}$$
$$= 1{,}901{,}413. \cdot 10^{-5} = 19.01413$$
$$\approx 19.0141 \text{ (after rounding to four decimal places),}$$

and it is seen that the columnar addition algorithm discussed in section 2.1.1 for the natural numbers works equally well for all decimal numbers. Thus the above (approximate) addition can be performed in the following concise format (all "carrying" being done mentally):

$$\begin{aligned} \frac{25}{8} &= 3.12500 \\ \frac{34}{3} &\approx 11.3333 \\ 2^{1/2} &\approx 1.41421 \\ \pi &\approx 3.14159 \end{aligned}$$

$$\frac{25}{8} + \frac{34}{3} + 2^{1/2} + \pi \approx 19.01413 \approx 19.0141.$$

The important consideration in using this algorithm (as in using it for the natural numbers) is to keep corresponding "places" (tens, units, tenths, hundredths, etc.) in the various summands vertically aligned. This can be done most easily by keeping the decimal points of all summands vertically aligned; the decimal point for the sum is then found in the same vertical alignment.

2.3.2. Multiplication

The decimal multiplication (or approximate multiplication) of real numbers is demonstrated in the following problem.

EXAMPLE

Computation of the product $(25/8) \cdot \pi$, using a five decimal place approximation for π.

Since $25/8 = 3.125$ and $\pi \approx 3.14159$ it is seen that

$$\begin{aligned}\frac{25}{8}\cdot\pi &\approx 3.125\cdot 3.14159\\ &= (3125.\cdot 10^{-3})\cdot(314{,}159.\cdot 10^{-5})\\ &= (3125.\cdot 314{,}159.)\cdot(10^{-3}\cdot 10^{-5})\\ &= (3125.\cdot 314{,}159.)\cdot 10^{-8}.\end{aligned}$$

The product of the integers 3125. and 314,159. can, of course, be computed by the algorithm of section 2.1.2. as follows:

```
     3 1 4 1 5 9.
         3 1 2 5.
     ------------
   1 5 7 0 7 9 5
   6 2 8 3 1 8
 3 1 4 1 5 9
9 4 2 4 7 7
-----------------
9 8 1 7 4 6 8 7 5.
```

The required product is then

$$\frac{25}{8}\cdot\pi \approx 981{,}746{,}875.\cdot 10^{-8} = 9.81746875.\dagger$$

Noting that the number of digits to the right of the decimal point in the final result is 8, or $3 + 5$, the sum of the number of digits to the right of the decimal point in multiplier and multiplicand respectively, it is seen that the entire computation could be presented in the form

† It should be noted that the value of the product as given here is an *approximation* to $25\pi/8$ and is based on the use of the *approximate* value 3.14159 for π. Since this latter value is not precisely correct, neither, of course, is the value of the product that is obtained and, as a matter of fact, several of the last digits in this product are erroneous. A more detailed explanation of such matters of accuracy is given on pages 244 thru 248.

$$
\begin{array}{r}
3.14159 \\
3.125 \\
\hline
1570795 \\
628318 \\
314159 \\
942477 \\
\hline
9.81746875,
\end{array}
$$

and that **the decimal point may be placed in the product by counting a number of decimal places equal to the sum of those in the multiplier and multiplicand.** Altho we shall not present any proof here, it is a fact that **this rule is valid in all cases.**

2.3.3. Subtraction

Since subtraction is the inverse of addition and the columnar algorithm of section 2.1.1 has been observed to work as well for decimal real numbers as for natural numbers, we suspect that the columnar algorithm of section 2.1.3 for subtraction of natural numbers may also work for the subtraction of decimal real numbers. That this is indeed the case is illustrated by the following example.

EXAMPLE

Computation of the difference $(25/8) - 2^{1/2}$, using a five decimal place approximation for $2^{1/2}$.

$$
\begin{aligned}
\frac{25}{8} - 2^{1/2} &\approx 3.125 - 1.41421 \\
&= (312{,}500. \cdot 10^{-5}) - (141{,}421. \cdot 10^{-5}) \\
&= (312{,}500. - 141{,}421.) \cdot 10^{-5} = 171{,}079. \cdot 10^{-5} \\
&= 1.71079.
\end{aligned}
$$

From the result it is clear that the vertical alignment algorithm of section 2.1.3 for natural numbers may be adapted to the present problem and the computation may be presented in the form:

$$\begin{array}{rcl} \frac{25}{8} & = & 3.12500 \\ 2^{1/2} & \approx & 1.41421 \\ \hline \frac{25}{8} - 2^{1/2} & \approx & 1.71079. \end{array}$$

2.3.4. Division

The quotient of any two terminating decimals (and hence the approximate quotient of any two real numbers) can always be represented as the quotient of two integers thus:

$$\frac{25/8}{2^{1/2}} \approx \frac{3.125}{1.41421} = \frac{3.125 \cdot 10^5}{1.41421 \cdot 10^5} = \frac{312{,}500.}{141{,}421.}.$$

Using the algorithm of section 2.1.4, we get

$$\begin{array}{r|l} & 2 \\ \hline 1\,4\,1\,4\,2\,1 & 3\,1\,2\,5\,0\,0 \\ & 2\,8\,2\,8\,4\,2 \\ \cline{2-2} & \,2\,9\,6\,5\,8 \end{array}$$

and hence

$$\frac{25/8}{2^{1/2}} \approx 2\,\frac{29{,}658}{141{,}421}.$$

If it is desired to convert the proper fraction 29,658/141,421 into decimal form, this can be done in the following manner as a continuation of the above division algorithm.

By writing a zero after the remainder 29,658., we convert it into the corresponding number, 296,580., of tenths. Since the divisor 141,421. may be subtracted from this number two times, we may represent this fact by placing a 2 in the "tenths place" of the quotient, that is, in the place immediately to the right of a decimal point placed to the right of the integral quotient thus:

```
                   2.2
        ______________
141421 | 312500.0
         282842
         ------
          296580
          282842
          ------
           13738.
```

After taking two times 141,421. from 296,580., a remainder of 13,738. (tenths) is left—that is, the actual remainder after taking *two tenths of 141,421.* from 29,658. is 1,373.8. This may be converted to 137,380. hundredths by adjoining a zero to the formal remainder of 13,738. Since this number (of hundredths) is less than the divisor, the latter can be subtracted from it zero times; hence the digit 0 is entered in the "hundredths place" of the quotient.

```
                   2.20
        ______________
141421 | 312500.00
         282842
         ------
          296580
          282842
          ------
           137380
           000000
           ------
           137380.
```

After subtracting 141,421. zero times from 137,380. we still have a remainder of 137,380. (hundredths) which we now convert to 1,373,800. thousandths by writing another zero after it. Since 141,421. may be subtracted from this latter number nine times, we so indicate by writing the digit 9 in the "thousandths place" of the quotient.

```
                 2.2 0 9
1 4 1 4 2 1.| 3 1 2 5 0 0.0 0 0
              2 8 2 8 4 2
              -----------
                2 9 6 5 8 0
                2 8 2 8 4 2
                -----------
                  1 3 7 3 8 0
                  0 0 0 0 0 0
                  -----------
                  1 3 7 3 8 0 0
                  1 2 7 2 7 8 9
                  -------------
                    1 0 1 0 1 1.
```

After subtracting 141,421. nine times from 1,373,800. thousandths (or subtracting nine thousandths of 141,421. from the actual remainder of 1,373,800) we now have a remainder of 101,011, or 101,011. thousandths, or, by writing another zero after it, 1,010,110. ten thousandths. Since 141,421. can be subtracted from this latter number seven times, we write the digit 7 in the "ten thousandths place" in the quotient, and subtract seven times 141,421. from 1,010,110. to get the next remainder of 2.0163, or 20,163. ten thousandths, or (by adjoining another zero) 201,630. hundred thousandths.

```
                 2.2 0 9 7
1 4 1 4 2 1.| 3 1 2 5 0 0.0 0 0 0
              2 8 2 8 4 2
              -----------
                2 9 6 5 8 0
                2 8 2 8 4 2
                -----------
                  1 3 7 3 8 0
                  0 0 0 0 0 0
                  -----------
                  1 3 7 3 8 0 0
                  1 2 7 2 7 8 9
                  -------------
                    1 0 1 0 1 1 0
                      9 8 9 9 4 7
                    -------------
                        2 0 1 6 3.
```

Since 141,421. can be subtracted from 201,630. one time, we obtain an entry of the digit 1 in the "hundred thousandths place" in the quotient. (Further entries in succeeding decimal places of the quotient could be obtained in a similar manner if desired.) Thus, at this stage, the use of our division algorithm (adapted for quotients containing decimal fractions as well as integers) yields

```
                    2.2 0 9 7 1
           _______________________
1 4 1 4 2 1.| 3 1 2 5 0 0.0 0 0 0 0
              2 8 2 8 4 2
              -----------
                2 9 6 5 8 0
                2 8 2 8 4 2
                -----------
                  1 3 7 3 8 0 0
                  1 2 7 2 7 8 9
                  -------------
                    1 0 1 0 1 1 0
                      9 8 9 9 4 7
                    -------------
                        2 0 1 6 3 0
                        1 4 1 4 2 1
                        -----------
                          6 0 2 0 9
```

and we notice that the number 60,209 is *less than half of the divisor* which justifies us in ignoring it if we wish to stop at the fifth decimal place. (If this last remainder had been greater than half of the divisor, then we should have had to round the quotient "up" to the next larger digit in the last decimal place to reflect such a fact properly.) Hence, finally,

$$\frac{25/8}{2^{1/2}} \approx 2.20971.$$

(Note: The contraction of the three-row entry

```
  1 3 7 3 8 0
  0 0 0 0 0 0
  -------------
  1 3 7 3 8 0 0
```

to the single entry

$$1\,3\,7\,3\,8\,0\,0$$

illustrates a legitimate short cut in the recording of divisional computation for instances where the digit 0 occurs in the quotient.)

A slightly different way of thinking thru essentially the same procedure that we have just used leads us to the following simple and more direct justification for our procedure. As before,

$$\frac{25/8}{2^{1/2}} \approx \frac{3.125}{1.41421} = \frac{3.125}{1.41421} \cdot \frac{10^5}{10^5} = \frac{312{,}500.}{141{,}421.}.$$

At this point we modify our thinking a bit from the previous pattern by multiplying both numerator and denominator of the fraction by 10^5, the choice of the *fifth* power of ten as the multiplier being dictated here by the fact that we wish to obtain the quotient accurately to the *fifth* decimal place. Thus we obtain

$$\frac{25/8}{2^{1/2}} \approx \frac{312{,}500.}{141{,}421.} = \frac{312{,}500.}{141{,}421.} \cdot \frac{10^5}{10^5} = \frac{31{,}250{,}000{,}000.}{141{,}421.} \cdot \frac{1}{10^5}$$
$$= \frac{31{,}250{,}000{,}000.}{141{,}421.} \cdot 10^{-5}.$$

Now by the use of the ordinary division algorithm for natural numbers the value of the quotient 31,250,000,000./141,412. can be obtained accurately to the nearest unit as 220,971. without any discussion of fractions, decimal or otherwise. Hence

$$\frac{25/8}{2^{1/2}} \approx 220{,}971. \cdot 10^{-5} = 2.20971.$$

2.3.5. Remarks on Precise Computation with Irrational Numbers

It will have been noticed that the discussion in the four preceding sections has been concerned mainly with the *approximate* computation of sums, products, differences and quotients of real numbers by substituting certain "approximately equal" terminating decimals in place of irrational numbers or of rational numbers with non-terminating decimal forms. A knowledge of this method is of practical importance because, while *some* computations even with irrational numbers can be performed *precisely*, it is impossible to perform precise computations with many irrational numbers and hence we must resort in such cases to our approximation methods as the only practical expedient. Following are some examples to illustrate both cases in which precise computations can be performed and those in which they cannot.

EXAMPLE 1

$$\begin{aligned}(45)^{1/2} + (20)^{1/2} &= (3^2 \cdot 5)^{1/2} + (2^2 \cdot 5)^{1/2}\\ &= 3 \cdot (5^{1/2}) + 2 \cdot (5^{1/2})\\ &= (3 + 2) \cdot (5^{1/2}) = 5 \cdot (5^{1/2}).\end{aligned}$$

EXAMPLE 2

$$\begin{aligned}(45)^{1/2} - (20)^{1/2} &= (3^2 \cdot 5)^{1/2} - (2^2 \cdot 5)^{1/2}\\ &= 3 \cdot (5^{1/2}) - 2 \cdot (5^{1/2})\\ &= (3 - 2) \cdot (5^{1/2}) = 1 \cdot (5^{1/2}) = 5^{1/2}.\end{aligned}$$

EXAMPLE 3

$$\begin{aligned}(45)^{1/2} \cdot (20)^{1/2} &= (45 \cdot 20)^{1/2} = [(3^2 \cdot 5) \cdot (2^2 \cdot 5)]^{1/2}\\ &= (2^2 \cdot 3^2 \cdot 5^2)^{1/2} = [(2 \cdot 3 \cdot 5)^2]^{1/2}\\ &= 2 \cdot 3 \cdot 5 = 30.\end{aligned}$$

EXAMPLE 4

$$\frac{(45)^{1/2}}{(20)^{1/2}} = \left(\frac{45}{20}\right)^{1/2} = \left(\frac{3^2 \cdot 5}{2^2 \cdot 5}\right)^{1/2} = \left(\frac{3^2}{2^2}\right)^{1/2} = \left[\left(\frac{3}{2}\right)^2\right]^{1/2} = \frac{3}{2}.$$

It will be noted that in each of the four preceding cases it was possible to obtain the *precise* value of the computed quantity—at least in indicated form—and no resort to decimal approximation was needed (altho such would be needed to express $5^{1/2}$ itself other than symbolically). In contrast to these cases we offer the following examples of expressions in which the indicated computations cannot be performed (even to the extent of symbolic simplification) without the use of decimal (or other) approximations.

EXAMPLES

$$2^{1/2} + \pi, \quad 2^{1/2} - \pi, \quad 2^{1/2} \cdot \pi, \quad 2^{1/2}/\pi.$$

EXERCISES

In those of the following problems which involve computations with square roots, make use of the approximate values given in the following table.

n	$n^{1/2}$
2	1.41421
3	1.73206
5	2.23607
6	2.44949
7	2.64575
8	2.82843
10	3.16228

1. Find, accurately to five decimal places, the decimal approximations for the following numbers.

(a) $\frac{154}{19}$ (b) $17\frac{6}{7}$ (c) $\frac{39}{9}$ (d) $\frac{43}{8}$ (e) $2\frac{17}{46}$ (f) $\frac{1}{11}$ (g) $\frac{1}{9}$ (h) $\frac{11}{13}$

(i) $\frac{1001}{101}$ (j) $\frac{1234}{4321}$.

2. Find, accurately to five decimal places, the approximate values of the following.

(a) $\frac{134}{17} + 3 \cdot (10^{1/2})$ (b) $5^{1/2} + 6^{1/2} + 7^{1/2}$ (c) $2 \cdot (3^{1/2}) - 3 \cdot (2^{1/2})$

(d) $2\frac{17}{46} + 10 \cdot (3^{1/2})$ (e) $\frac{1}{11} + \frac{1}{9} + \frac{1234}{4321} + 2^{/12}$.

3. Find, accurately to five decimal places, the approximate values of the following.

(a) $2^{1/2} + 3^{1/2} - 5^{1/2}$ (b) $7^{1/2} - 6^{1/2}$ (c) $8^{1/2} - 2 \cdot (2^{1/2})$

(d) $(2^{1/2}) \cdot (3^{1/2}) - 6^{1/2}$ (e) $\frac{1234}{4321} - \frac{1001}{101}$.

4. Find accurately to five decimal places, the approximate values of the following.

(a) $(3^{1/2}) \cdot (5^{1/2})$ (b) $(2^{1/2}) \cdot (8^{1/2})$ (c) $(3^{1/2}) \cdot (5^{1/2}) \cdot (7^{1/2})$

(d) $(5^{1/2}) \cdot (6^{1/2}) - (3^{1/2}) \cdot (10^{1/2})$ (e) $5 \cdot (6^{1/2}) - 6 \cdot (5^{1/2})$.

5. Find, accurately to five decimal places, the approximate values of the following.

(a) $\dfrac{8^{1/2}}{2^{1/2}}$ (b) $\dfrac{3^{1/2}}{17\frac{6}{7}}$ (c) $\left(\dfrac{5^{1/2}}{3^{1/2}}\right) - \left(\dfrac{10^{1/2}}{6^{1/2}}\right)$ (d) $\dfrac{1/11 - 1/9}{2^{1/2}}$

(e) $\dfrac{7^{1/2}}{\frac{43}{8} - 2\frac{17}{46}}$.

6. Find, accurately to five decimal places, the value of

$$\frac{(m+n)x+y}{z}$$

in each of the following cases. Explain any exceptional case.

(a) $m = 2^{1/2}$ $n = 3^{1/2}$ $x = 5^{1/2}$ $y = 6^{1/2}$ $z = 7^{1/2}$

(b) $m = \dfrac{154}{19}$ $n = 17\dfrac{6}{7}$ $x = \dfrac{39}{9}$ $y = \dfrac{43}{8}$ $z = 2\dfrac{17}{46}$

(c) $m = \dfrac{1}{11}$ $n = 8^{1/2}$ $x = \dfrac{1}{9}$ $y = 10^{1/2}$ $z = \dfrac{11}{13}$

(d) $m = 3^{1/2}$ $n = 2\dfrac{17}{46}$ $x = \dfrac{1001}{101}$ $y = 7^{1/2}$ $z = 21$

(e) $m = \dfrac{39}{9}$ $n = 10^{1/2}$ $x = \dfrac{1}{11}$ $y = \dfrac{1}{9}$ $z = 0$.

7. If *only* the values for $2^{1/2}$, $3^{1/2}$ and $5^{1/2}$ had been given in the table preceding the present set of exercises, show how each of the following numbers could have been obtained (without using any special procedure for computing square roots): (a) $6^{1/2}$; (b) $8^{1/2}$; (c) $10^{1/2}$.

(d) Explain why $4^{1/2}$ and $9^{1/2}$ were not included in the table.

2.3.6. Square Root Extraction

The problem of "extracting" the positive square root of a positive real number r is the problem of finding the positive number s for which

$$s^2 = r.$$

Since we know that the area of a square is the product obtained by multiplying the length of an edge of the square by itself, we can think of square root extraction as the problem of finding the length s of the edge of a square whose area is a given number r.

Like most computational processes, square root extraction may start with trial-and-error type experimentation. The usual

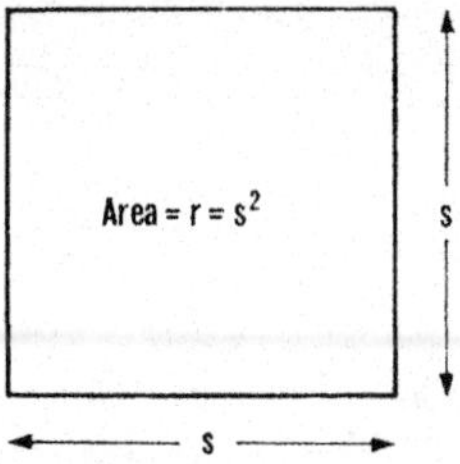

historical course of development from trial-and-error procedure to a streamlined and efficient algorithm is characterized by the reduction of both the number of trials and the number and magnitude of the consequent errors to a minimum. We shall illustrate this evolutionary process in the discussion of the following problem.

Let it be required to find the length s (in feet) of an edge of a square whose area is 614,656 square feet.

Since a square of length 1000 ft would have an area of 1000^2 or 1,000,000 sq. ft, it is obvious that the number s which we are seeking must be less than 1000. Similarly, it is clear that s must be greater than 500 since $500^2 = 250{,}000$ which is much smaller than the given area. In the same way, since $800^2 = 640{,}000$ and $700^2 = 490{,}000$, we find that s is less than 800 but more than 700.

Another way of expressing our findings in the preceding paragraph is to say that s is 700 plus some unknown amount, which we may think of as a "correction" to be added to 700 to get the correct value of s, the value of the correction being less than 100. Our original problem will be solved by finding the value of this correction.

If the desired correction were exactly x feet then we should have

$$(700 + x)^2 = 614{,}656$$

or

$$700^2 + 2 \cdot 700 \cdot x + x^2 = 614{,}656$$

whence

$$2 \cdot 700 \cdot x + x^2 = 614{,}656 - 490{,}000$$

or

$$(2 \cdot 700 + x) \cdot x = 124{,}656$$

and

$$x = \frac{124{,}656}{2 \cdot 700 + x}.$$

Since x is less than 100 and hence *small* as compared with the term $2 \cdot 700$ we may replace the denominator $2 \cdot 700 + x$ by the roughly approximately equal expression $2 \cdot 700$ to obtain the roughly approximate value

$$x \approx \frac{124{,}656}{2 \cdot 700} = \frac{124{,}656}{1400} \approx 89.$$

However, since this value is obtained by replacing the exact denominator $2 \cdot 700 + x$ by a smaller one, it is seen that the value 89 must be larger than the exact value of x for which the relation

$$x = \frac{124{,}656}{2 \cdot 700 + x}$$

would be true, and hence that this exact value of x must be somewhat smaller than 89.

For the sake of the convenience of working with round numbers let us try a correction of 80 and see if such a value of x would be adequate. Our first estimate of 700 accounted for 700^2 or 490,000 of the given 614,656 sq. ft of area, leaving 641,656 − 490,000 or 124,656 sq. ft unaccounted for. A correction of 80 would account for an additional area of

$$2 \cdot 700 \cdot 80 + 80^2, \text{ or } 118,400, \text{ sq. ft}$$

leaving

$$124{,}656 - 118{,}400, \text{ or } 6256, \text{ sq. ft}$$

still unaccounted for. (See figure on next page.)

Since after applying a correction of 80 to get an improved estimate of 700 + 80 or 780 for s, we still have 6256 sq. ft of the given area unaccounted for, it is obvious that the correction of 80

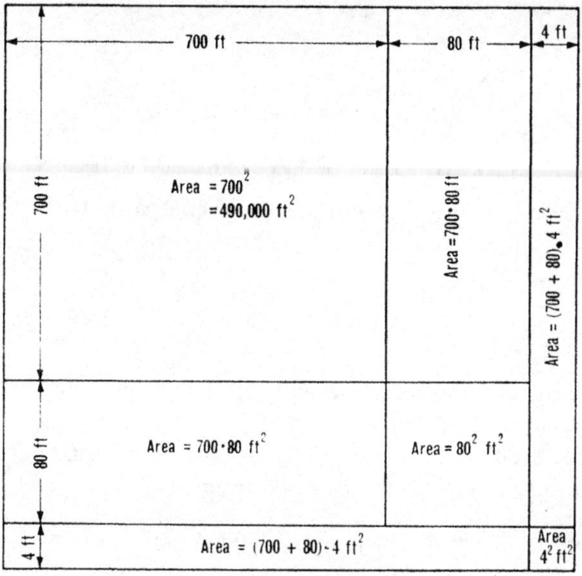

was not quite large enough. Hence an additional correction is now needed. Let us call this required additional correction y. Then we should have

$$[(700 + 80) + y]^2 = 614{,}656$$

or

$$(700 + 80)^2 + 2 \cdot (700 + 80) \cdot y + y^2 = 614{,}656$$

or

$$700^2 + 2 \cdot 700 \cdot 80 + 80^2 + 2 \cdot (700 + 80) \cdot y + y^2 = 614{,}656$$

or

$$700^2 + (2 \cdot 700 + 80) \cdot 80 + [2 \cdot (700 + 80) + y] \cdot y = 614{,}656$$

whence

$$\begin{aligned}[2 \cdot (700 + 80) + y] \cdot y &= 614{,}656 - 700^2 - (2 \cdot 700 + 80) \cdot 80 \\ &= 124{,}656 - (2 \cdot 700 + 80) \cdot 80 \\ &= 6256\end{aligned}$$

and

$$y = \frac{6256}{2 \cdot (700 + 80) + y}$$

so that

$$y \approx \frac{6256}{2 \cdot (700 + 80)}, \text{ since } y \text{ is small compared with } 2 \cdot (700 + 80),$$

$$= \frac{6256}{1560} \approx 4.$$

Again, our tentatively obtained value of y may be too large since we got it by dropping the "small" term y from the expression $2 \cdot (700 + 80) + y$. Testing the value 4 for the additional correction y, we find that if $y = 4$, then

$$[2 \cdot (700 + 80) + y] \cdot y = [2 \cdot (700 + 80) + 4] \cdot 4$$
$$= 1564 \cdot 4 = 6256$$

which exactly accounts for the remaining area of the originally given 614,656 sq. ft.

Thus the required value for the edge-length s is

$$s = 700 + 80 + 4 = 784 \text{ ft}$$

since $784^2 = 614{,}656$.

We have solved our problem completely and precisely by making tentative estimates (or "guesses") and then applying successive corrections to them. The only thought which lingers to dim our feeling of jubilation is that it took us several pages to do it. With our motto "There must be an easier way" to guide and inspire us, let us see how concisely and economically we can record the essential numbers involved in the computation.

First, there is the number **614,565** of square feet in the given area. This number is an essential characteristic of our problem and must be written down for reference.

Second, we subtracted 490,000 sq. ft from the given area to account for our first estimate of 700 ft for the edge-length s. The number **700**, of course, should be recorded and the number **490,000** must be written in some manner to indicate the role it plays in the computation.

Third, we deduced an approximate correction of 80 and added this to our previous estimate of 700 to get an improved estimate of 780 ft for s. Also, we further subtracted from what remained of the original area, an additional amount (two rectangular areas of $700 \cdot 80$ sq. ft each and a small square area of 80^2 sq. ft) corresponding to our enlargement of the estimate for s. Here the quantities to be written are the correction **80** and the additional area subtracted, namely, $2 \cdot 700 \cdot 80 + 80^2$, or $\mathbf{(2 \cdot 700 + 80) \cdot 80}$, or **118,400** sq. ft.

Fourth, we deduced an additional correction of 4 to the previous value of s and then, corresponding to this enlargement of the estimate for s, subtracted $[2 \cdot (700 + 80) + 4] \cdot 4$ which exhausted all that remained of the original area, showing that our final corrected value of s was precisely the value we sought. The numbers **4** and $\mathbf{[2 \cdot (700 + 80) + 4] \cdot 4}$, or **6256**, are essential to the computation and hence must be written.

The information discussed in the four preceding steps may be arranged in the following manner.

Length of edge

Given area ⟶ 6 1 4 6 5 6 | 7 0 0 + 8 0 + 4

Subtracted area of 700^2 ⟶ 4 9 0 0 0 0

Approximate divisor,

$2 \cdot 700$ ⟶ 1 4 0 0 | 1 2 4 5 6 5 ← Remaining area

1 1 2 0 0 0 } Area subtracted corresponding to first correction

6 4 0 0 }

Approximate divisor,

```
2 · (700 + 80) → 1560 | 6256 ←Remaining area
                        6240 ⎫ Area subtracted
                             ⎬ corresponding to
                          16 ⎭ second correction
                        ----
Remaining area ————————————→ 0.
```

Some further simplication can be obtained by taking advantage of the principle of positional notation to write 49 for 490,000, 1184 for 118,400, etc., and also by writing the computed square root one digit at a time. Thus the essential numerical record of the entire computation may be presented in the following condensed format.

```
         614656 | 784
         49
         --
   148 | 1246
         1184
         ----
  1564 | 6256
         6256
         ----
            0.
```

Important practical note. It is to be noted that the "hundreds "digit" 7, (the *third* digit to the left of the decimal point in the computed square root 784) is the largest digit whose square does not exceed the two-digit number 61 (the *third* such two-digit grouping to the left of the decimal point in the given number 614,656). **The leftmost digit in any square root is always obtained in an analogous manner.** When the 700^2, or 490,000, is subtracted from 614,656, with 490,000 represented by the abbreviation "49", this 49 is subtracted from the "61" in 614,656.

The first correction 80 is obtained as the approximate quotient of the remainder 124,656 divided by the "trial divisor" $2 \cdot 700$, or 1400. When, in the condensed algorithm, the trial divisor 1400 is represented by "14" and the correction 80 by "8", the remainder 124,656 must be correspondingly modified. Since

$$\frac{124{,}656}{1400} \approx 80$$

or

$$124{,}656 \approx 1400 \cdot 80 = (14 \cdot 10^2) \cdot (8 \cdot 10) = (14 \cdot 8) \cdot 1000,$$

it follows that

$$14 \cdot 8 \approx \frac{124{,}656}{1000} \approx 124$$

or

$$\frac{124}{14} \approx 8.$$

That is, the 8 (representing the correction 80) is obtained as the approximate quotient of *124* (representing the remainder 124,656) divided by the trial divisor 14 (representing 1400).

Similarly, the second correction 4 is obtained as the approximate quotient of *625* (representing the remainder 6256) divided by the trial divisor 156 (representing 1560).

In obtaining each successive correction, **the trial divisor is double the sum of the first square root conjecture plus all previous corrections. The number into which this trial divisor is divided** (in order to obtain the next correction) **is the result of the last subtraction with just the first one (not both) of the next two digits "brought down" from the given number adjoined.**

The following example illustrates how, when a given number is not the exact square of any integer, its approximate square root may be computed accurately to as many decimal places as may be desired.

EXAMPLE

Approximate computation of $614{,}953^{1/2}$.

```
                 6 1 4 9 5 3.0 0 0 0 0 0 0 0 ... | 7 8 4.1 8 9 3 ...
                 4 9
                 ---
       1 4 8 |   1 2 4 9
                 1 1 8 4
                 -------
         1 5 6 4 |   6 5 5 3
                     6 2 5 6
                     -------
         1 5 6 8 1 |   2 9 7 0 0
                       1 5 6 8 1
                       ---------
       1 5 6 8 2 8 |   1 4 0 1 9 0 0
                       1 2 5 4 6 2 4
                       -------------
       1 5 6 8 3 6 9 |   1 4 7 2 7 6 0 0
                         1 4 1 1 5 3 2 1
                         ---------------
         1 5 6 8 3 7 8 3 |   6 1 2 2 7 9 0 0
                             4 7 0 5 1 3 4 9
                             ---------------
                             1 4 1 7 6 5 5 1.
```

Once learned, this square root algorithm adds considerably to one's personal computational resources and independence of tables, slide rules and calculating machines. Its acquisition should be a source of satisfaction to the challenge-seeking student and a sharpener of his computational skills; it may even serve to stimulate his interest in numbers and their intrinsic properties, and his ingenuity in developing original algorithms of his own. While it is admitted that the ability to compute square roots "by hand" is certainly not a necessity for every child, it is believed that every child should have the opportunity to see this highly instructive algorithm demonstrated and have its working explained if his interest prompts him to seek such explanation.

2.3.7. Rounding Off; Significant Digits

When we speak of the irrational number π as being "approximately 3.1416", we say that we are "rounding it off" and, in particular, that we are "rounding it off to the fourth decimal place". We might, if its value were not required to be used with great accuracy, further round it off to 3.14, or even to 3.

Similarly, the rational number 1/3 which has the non-terminating decimal representation .33333 ... may, for convenience, be rounded off to the approximate value .33333, or to .33, or even to .3.

Sometimes, when exactness is not considered important, even a terminating decimal like 3.625 might be rounded off to a simpler (but only approximate) value such as 3.6.

The extent to which it may be permissible to round off numbers in this manner depends on the accuracy with which the numbers involved (and others obtained from them by computation) are required to be known. For instance, in using the number π in computations involving the orbit of a space vehicle, it might be necessary to use a decimal approximation accurate to ten decimal places, whereas in computing the circumference of a circular garden plot it might be quite satisfactory to use the value 3.1 as an approximation to π.

An important consideration in rounding off numbers is to be *consistent.* For example, in measuring the three sides of a triangle to determine its perimeter (as their sum) it would be pointless to determine one side with great accuracy (say, as 7.36945 ft), a second with moderate accuracy (say, as 4.57 ft), and the third quite crudely (say, as "about 6 ft"), for, after adding them, the sum obtained would be no more accurate than the least accurate of the three sides. To form the sum as

$$\begin{array}{l} 7.3\,6\,9\,4\,5 \\ 4.5\,7 \\ 6. \\ \hline 17.9\,3\,9\,4\,5 \end{array}$$

would be to give an impression of great accuracy in the sum which would be quite false and misleading. Since the second length might actually be anything between 4.56500 ft and 4.57500 ft and the third length anything between 5.50000 ft and 6.50000 ft, the actual perimeter might be anything between 17.43445 ft and 18.44445 ft. Hence we round off the sum to match the accuracy of the least accurate of the given data and report the perimeter of the triangle (crudely but honestly) to be "about 18 ft".

From the above example it may be seen that when numbers are to be added or subtracted, they should all be determined (or rounded off) with accuracy to the same number of decimal places if possible.

In multiplication a somewhat different principle must be kept in mind. This principle is illustrated by the following example.

Suppose that a number L is 231.684 and a second number W is 8.357. Then the product $L \cdot W$ is 1936.183188. If L is the length of a rectangle and W is its width and if L is *precisely* 231.684 inches (that is, $L = 231.684000 \ldots$) and W is *precisely* 8.357000 ... then, of course, the area $L \cdot W$ of the rectangle is *precisely* 1936.183188000 ... square inches. On the other hand, if the values 231.684 and 8.357 are only *approximations* to the actual or precise values of L and W respectively, then the number 1936.183188 is only an *approximation* to the actual area $L \cdot W$.

Every real number has the decimal form

$$I\,.\,d_1d_2d_3d_4d_5d_6\,\ldots,$$

where I is the integral part and the "decimal digits" d_1, d_2, d_3, etc., are respectively the tenths digit (that is, the number of tenths), the hundredths digit, the thousandths digit, etc. Hence we may suppose the exact values of L and W to be of the form

$$L = I_L \,.\, L_1\, L_2\, L_3\, L_4\, L_5\, L_6 \ldots$$
$$W = I_W \,.\, W_1\, W_2\, W_3\, W_4\, W_5\, W_6 \ldots .$$

If 231.684 is only an approximation to L, this means that it is *known* that $I_L = 231$, $L_1 = 6$, $L_2 = 8$, and $L_3 = 4$† but that L_4, L_5, L_6, and all the rest of the decimal digits in L are *unknown*. Likewise, if 8.357 is only an approximation to W, then it is *known* that $I_W = 8$, $W_1 = 3$, $W_2 = 5$ and $W_3 = 7$ but W_4, W_5, W_6, and all the rest of the decimal digits in W are *unknown*. Now let us examine the degree of certainty of our resulting knowledge of the product $L \cdot W$.

$$\begin{aligned} L \cdot W &= (I_L + L_1 \cdot 10^{-1} + L_2 \cdot 10^{-2} + L_3 \cdot 10^{-3} + L_4 \cdot 10^{-4} \\ &\quad + L_5 \cdot 10^{-5} + L_6 \cdot 10^{-6} + \ldots) \cdot (I_W + W_1 \cdot 10^{-1} + \\ &\quad W_2 \cdot 10^{-2} + W_3 \cdot 10^{-3} + W_4 \cdot 10^{-4} + W_5 \cdot 10^{-5} + \\ &\quad W_6 \cdot 10^{-6} + \ldots) \\ &= (231.684 + L_4 \cdot 10^{-4} + L_5 \cdot 10^{-5} + \ldots) \\ &\qquad \cdot (8.357 + W_4 \cdot 10^{-4} + W_5 \cdot 10^{-5} + \ldots) \\ &= 231.684 \cdot 8.357 + 231.684 \cdot (W_4 \cdot 10^{-4} + W_5 \cdot 10^{-5} \\ &\qquad + \ldots) \\ &\quad + 8.357 \cdot (L_4 \cdot 10^{-4} + L_5 \cdot 10^{-5} + \ldots) \\ &\quad + (L_4 \cdot 10^{-4} + L_5 \cdot 10^{-5} + \ldots) \\ &\qquad \cdot (W_4 \cdot 10^{-4} + W_5 \cdot 10^{-5} + \ldots) \\ &= 1936.183188 + 231.684 \cdot W_4 \cdot 10^{-4} + \text{smaller terms.} \end{aligned}$$

Now noticing that $231.684 \cdot W_4 \cdot 10^{-4} = .0231684 \cdot W_4$ and that (depending on the value of W_4) this may not only be greater than .02 but possibly even nearly .1, it is clear that the value of this term alone may affect the value of $L \cdot W$ in the first decimal place and thus that the computed result 1936.183188 may not be

† Actually there is a slight over-simplification involved here, but to go into great detail concerning the exceptional finer points might tend to obscure the main point we wish to make.

accurate at all beyond the first four digits. Thus in this case we can only say that $L \cdot W$ is about 1936, and we are not justified in making any statement concerning the value of $L \cdot W$ which involves greater accuracy than this.

Each of the digits in the number 8.357 is called **significant** because its written presence is assumed to signify actual accurate knowledge of W as far as the units, tenths, hundredths and thousandths places are concerned. Similarly, each of the digits in the number 231.648 is called significant because it is assumed to signify factual knowledge about L as determined by reliable measurement of some sort. If these numbers were to be rewritten as 08.357, 008.357, 0231.685, etc., the zeros on the left would not be considered as significant digits because their presence would not signify any additional knowledge concerning the actual values of L and W. Similarly, zeros written to the right, as in 8.3570, 8.35700, etc., would be considered significant *only* if their presence signified actual definite knowledge concerning the value of W as it would affect the indicated decimal places. A zero occurring between two non-zero digits, as in 8.057, would always be significant since a zero would never be written in such a position except when its presence was intended to indicate actual numerical knowledge concerning the value of the quantity being measured.

Since the concept of significant digits is both useful and important in discussing the accuracy of numerical computations and their results, we shall use the following as a standard practical working definition of the concept. **In any decimal (or other positional notation) representation of a number, those digits appearing in its representation which are called significant are: the left-most non-zero digit and all other digits written to the right of this whose presence is known to signify actual knowledge of the represented number in their respective places.**

EXAMPLES

1. If a number is written in scientific notation as $6.00 \cdot 10^8$, then only the 6 and the two zeros written immediately following are considered as significant. If this number is rewritten in the

form 600,000,000, the first three digits are still the only ones which are considered as significant.

2. In the number .00003 only the digit 3 is considered as significant. If converted to scientific notation, it would be written as $3 \cdot 10^{-5}$.

3. A number would be written as .000030 only if the last zero reflected actual knowledge (in the sixth decimal place) of the represented quantity. In scientific notation such a number would appear as $3.0 \cdot 10^{-5}$.

4. In the number 100.00001 all written digits are considered significant.

5. In the number .00010001 only the two ones and the three zeros *between* them are considered significant.

It may be noted that in the multiplication problem which we discussed, there are *four* significant digits in the value 8.357 which we assumed to be accurately known for W, *six* significant digits in the value 231.684 which we assumed to be accurately known for L, and only *four* significant digits that we could deduce to be *accurately* known in the resulting product. This exemplifies the following general rule.†

The number of significant digits in a computed product is not greater than the least number of significant digits which occurs in any of its factors. This rule should be kept in mind in connection with all problems involving the multiplication of decimal numbers. The same rule holds effectively for the accuracy of computed quotients.

We may summarize our findings on accuracy of numerically computed results as follows. **In addition and subtraction, it is the rightmost accurately known digit in the least accurately known term which determines the rightmost accurate digit in the computed sum or difference. In multiplication and division, it is the number of significant digits in the least accurately known factor which deter-**

† The rule stated here is not precise in the sense that there is no exception to it. It is a rough but satisfactory working rule in the sense that exceptions to it are encountered relatively infrequently and such exceptions are usually not of serious importance.

mines the number of significant digits in the computed product or quotient.

EXERCISES

1. Compute accurately to five decimal places and tabulate the values of the square roots of all prime numbers not greater than 20.

2. Explain how the square root of any composite number not greater than 40 can be computed indirectly by using the results of Exercise 1. As examples, use this method to compute $27^{1/2}$, $28^{1/2}$ and $33^{1/2}$.

3. Find and explain a way to obtain the value of $(1/n)^{1/2}$ by computing $n^{1/2}$, where n is a positive integer.

4. The following method is sometimes suggested for computing square roots. To find the square root of a positive number n, start by "guessing" any positive number, say r_1. Then divide n by r_1 to obtain a quotient q_1 which is then averaged† with r_1 to obtain a number r_2. This number r_2 is then treated as a second guess, or approximation, to the desired square root. To get a third approximation r_3, n is divided by r_2 to get a quotient q_2 which is then averaged with r_2 to produce r_3. This process is repeated to obtain successive approximations r_4, r_5, r_6, etc., the process terminating when two successive approximations agree at the desired level of accuracy. Compute the square root of 2954.8264 accurately to four significant digits both by this method and by the standard algorithm. Show a complete but concise record of all your numerical computations in each case and then state your conclusion about the relative efficiency of the two methods.

5. Compute the square root of each of the following accurately to five significant digits.

(a) 26,593.6805 (b) 3/8 (c) 235,000,000,000 (d) .6284000000
(e) .0000000432000008.

6. (a) Deduce a relation between the number of significant digits in a given number and the number of significant digits in its computed square root. (b) Deduce a relation between the number of significant digits in a given number and the number of significant digits in its computed square.

7. Perform the indicated computations. In each case consider the data given, not as representing precisely determined values, but as representing approximations which are accurate only to the number of significant digits which appear. Round off the computed result in a manner consistent with the accuracy of the given data.

(a) $23.4621 + 218.197 + 1.06421 - 2392.20$

(b) $(22.89) \cdot (57.95) - 33.47$

(c) $\dfrac{12.027 - 11.998}{21{,}392.}$ (d) $\dfrac{1.0014}{712. - 698.}$

(e) $(.08)\,(.3)\,(1.02)\,(.77) - 0000004002.$

† Here the "average" of r_1 and q_1 is to be understood as the number $(r_1 + q_1)/2$.

CHAPTER 2.4

COMPUTATION WITH THE COMPLEX NUMBERS

2.4.1. Numerical Computation

It has been seen in section 1.4.5.2 that, for any complex numbers $a + bi$ and $c + di$, their sum, difference, product and quotient are given by the relations:

$$(a + bi) + (c + di) = (a + c) + (b + d)i$$

$$(a + bi) - (c + di) = (a - c) + (b - d)i$$

$$(a + bi) \cdot (c + di) = (ac - bd) + (ad + bc)i$$

$$\frac{a + bi}{c + di} = \frac{ac + bd}{c^2 + d^2} + \frac{bc - ad}{c^2 + d^2}i.$$

Since these relations may be used as formulas for performing the four fundamental operations of arithmetic with any given complex numbers, this completely solves the problem of all such arithmetical computations.

EXAMPLES

1. $(71 + 69i) + (12 - 5i) = (71 + 12) + (69 + (-5))i$
$= 83 + 64i.$

2. $(71 + 69i) - (12 - 5i) = (71 - 12) + (69 - (-5))i$
$= 59 + 74i.$

3. $(71 + 69i) \cdot (12 - 5i) = (71 \cdot 12 - 69 \cdot (-5)) + (71 \cdot (-5) + 69 \cdot 12)i$
$= (852 + 345) + (-355 + 828)i$
$= 1197 + 473i.$

4. $$\frac{71+69i}{12-5i} = \frac{71\cdot 12+69\cdot(-5)}{12^2+(-5)^2} + \frac{69\cdot 12-71\cdot(-5)}{12^2+(-5)^2}\,i$$

$$= \frac{852-345}{144+25} + \frac{828+355}{144+25}\,i = \frac{507}{169} + \frac{1183}{169}\,i$$

$$= 3+7i.$$

5. $(.347 - .962i) + (-.561 + .279i)$

$$= (.347 + (-.561)) + (-.962 + .279)i$$

$$= -.214 - .683i.$$

6. $$\frac{.347 - .962i}{-.561 + .279i} = \frac{(.347)\cdot(-.561) + (-.962)\cdot(.279)}{(-.561)^2 + (.279)^2}$$

$$+ \frac{(-.962)\cdot(-.561) - (.347)\cdot(.279)}{(-.561)^2 + (.279)^2}\,i$$

$$= -1.18 + 1.13i.$$

2.4.2. Graphical Representation of Complex Sums, Differences, Products, Quotients, Powers and Roots

2.4.2.1. *Graphical addition*

If, as on page 137, we represent the complex numbers $a + bi$ and $c + di$ by points P and Q with the respective coordinates (a,b) and (c,d), then, of course, the sum $(a + bi) + (c + di)$, which is equal to $(a + c) + (b + d)i$, would likewise be represented by the point S with coordinates $(a + c,\ b + d)$.

From the geometry of the figure it is seen that the points O, P, Q and S form a parallelogram with P and Q as one pair of opposite vertices and O and S as the other pair. This observation enables us to formulate the following rule for adding two complex numbers graphically.

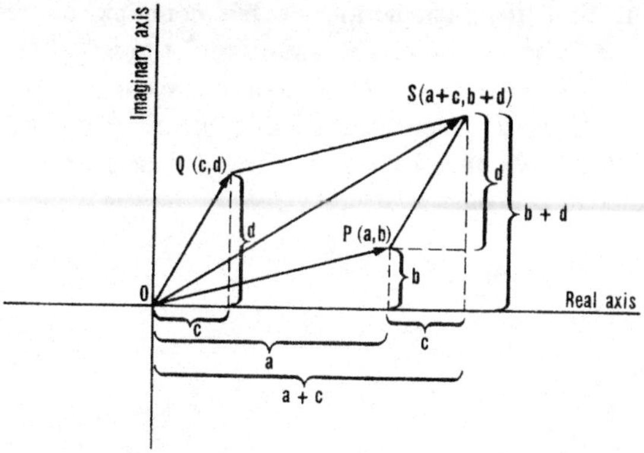

To find the sum, graphically, of the complex numbers $a + bi$ and $c + di$, first locate the point P with coordinates (a,b) to represent $a + bi$ and the point Q with coordinates (c,d) to represent $c + di$, then complete the parallelogram $OPSQ$ with OP and OQ as adjacent sides. The fourth vertex S of this parallelogram will then have coordinates $(a + c, b + d)$ and will represent the sum $(a + bi) + (c + di)$.

2.4.2.2. *Graphical subtraction*

The additive identity element for the complex numbers is the number $0 + 0 \cdot i$ which, for all practical purposes, is the same as the real number 0.

The additive inverse of the complex number $a + bi$ is defined, in the usual manner, as that unique complex number which when combined in addition with $a + bi$ (in either order) produces a sum equal to the additive identity element, zero. We represent this additive inverse symbolically as $-(a + bi)$ and call it "the negative of $a + bi$". From the definition it is seen that it is equal to $-a - bi$. It should be noted that the word "negative" is used here in a *relative* sense only, since individual complex numbers in general are not categorized as either positive or negative.

Since

$$(a + bi) - (c + di) = (a + bi) + [-(c + di)]$$

it is seen that **the subtraction of the complex number** $c + di$ **from** $a + bi$ **is the same as the addition of the negative of** $c + di$ **to** $a + bi$. Hence **the graphical determination of the difference** $(a + bi) - (c + di)$ **is effected by first locating on the graph the negative of** $c + di$**, that is,** $-c - di$**, and then graphically adding this number to** $a + bi$.

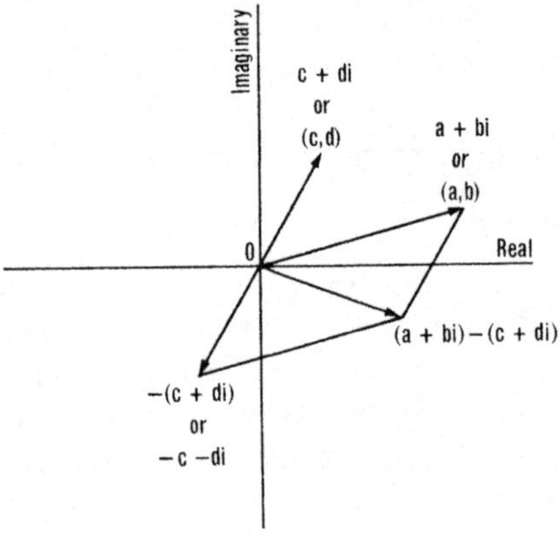

2.4.2.3. *Graphical multiplication*

A directed line segment drawn from one point to another (such as *OP*, *OQ* or *OS* in the figure on page 251) is called a **vector.** The **angle of inclination** of a vector is that angle measured positively in a counterclockwise direction (and not to exceed 360°) from the right-hand end of a horizontal line (such as the real axis) to the vector.

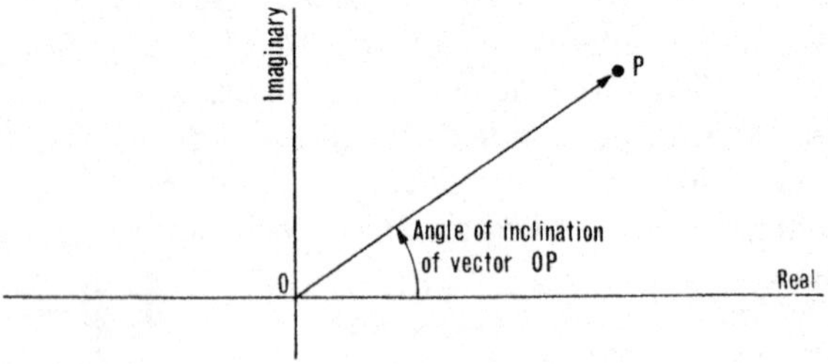

If the complex numbers $a + bi$ and $c + di$ and their product $(a + bi) \cdot (c + di)$, or $(ac - bd) + (ad + bc)i$, be represented by the points P, Q and R respectively and the complex number 1 (or $1 + 0 \cdot i$) be represented by the point U, then it can be shown that the triangles OUP and OQR are *similar*, and from this it can be shown that the length of the vector OR is equal to the product of the lengths of the vectors OP and OQ, and that the angle of inclination of OR is equal to the sum of the angles of inclination of OP and OQ. This leads to the following rule for the graphical multiplication of complex numbers.

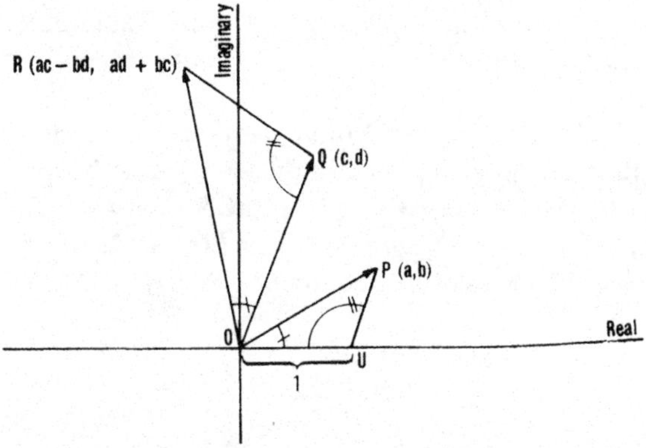

To determine graphically the product of the complex numbers $a + bi$ and $c + di$ first locate the points $U(1,0)$, $P(a,b)$ and $Q(c,d)$; then with OQ and OU as corresponding sides, construct the triangle OQR similar to the triangle OUP. The point R will then have coordinates $(ac - bd, ad + bc)$ and will represent the product $(a + bi) \cdot (c + di)$.

2.4.2.4. *Graphical division*

The figure just shown to illustrate graphical multiplication will also serve for graphical division if the points P, Q and R be reinterpreted. Since to divide $u + vi$ by $x + yi$ is to find a number $w + zi$ such that

$$u + vi = (x + yi) \cdot (w + zi),$$

we may find the required quotient in the following way. **Locate a point R to represent $u + vi$ and a point Q to represent $x + yi$. If a triangle OUP is then constructed similar to OQR, the point P will have coordinates (w,z) and will represent the quotient $w + zi$, or $(u + vi)/(x + yi)$.**

2.4.2.5. *Graphical construction of integral powers*

Since any power whose exponent is a positive integer is just an extended product with all nontrivial factors equal, the rule given for graphical multiplication of complex numbers can be adapted to the determination of such positive integral powers of a complex number.

In particular, **if the point $N(a,b)$ represents the complex number $a + bi$, then the point representing $(a + bi)^2$ and hence having coordinates $(a^2 - b^2, 2ab)$ will be a point S such that the length of OS is the square of the length of ON and the angle of inclination of OS is twice the angle of inclination of ON.**

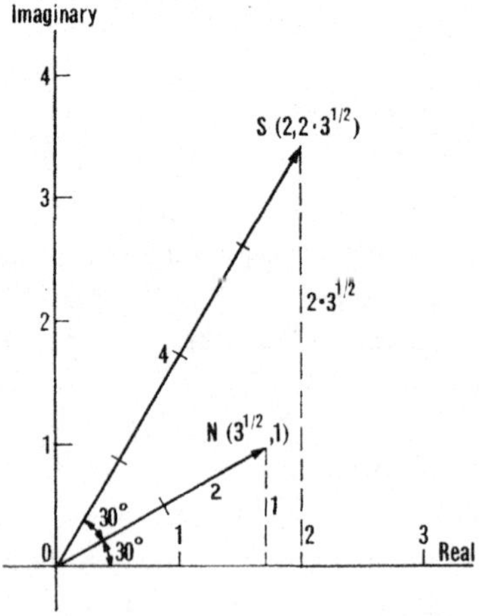

EXAMPLE

$$(3^{1/2} + i)^2 = 3 + 2 \cdot 3^{1/2} \cdot i + i^2$$
$$= 3 + 2 \cdot 3^{1/2} \cdot i - 1$$
$$= 2 + 2 \cdot 3^{1/2} \cdot i.$$

In general, if the point $N(a,b)$ represents the complex number $a + bi$, then the point which represents $(a + bi)^m$ is a point S such that the length of OS is the mth power of the length of ON and the angle of inclination of OS is m times the angle of inclination of ON.

EXAMPLE

$$(1 + i)^3 = (1 + i)^2 \cdot (1 + i) = (1 + 2i + i^2) \cdot (1 + i)$$
$$= (1 + 2i - 1) \cdot (1 + i) = 2i \cdot (1 + i)$$
$$= 2i + 2i^2 = 2i - 2$$
$$= -2 + 2i.$$

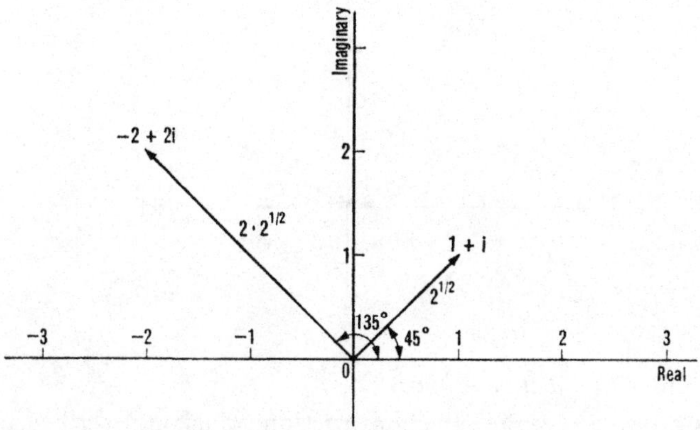

A particularly interesting special case involves the graphical distribution of the various positive integral powers of a complex number such that the length of the vector drawn from the origin to the point representing the number is 1. In this case the length of the vector drawn from the origin to the point representing the mth power of the number is 1^m which, of course, is also 1 regardless of the value of m. Hence the points which represent powers of such a number all lie on a circle with center at the origin and radius 1.

EXAMPLE

Consider the complex number $z = \frac{1}{2}(3^{1/2}) + \frac{1}{2}i$. The length of the vector drawn from the origin to the point representing z is

$$\{[\tfrac{1}{2}(3^{1/2})]^2 + (\tfrac{1}{2})^2\}^{1/2} = \{\tfrac{3}{4} + \tfrac{1}{4}\}^{1/2} = 1^{1/2} = 1.$$

Hence all powers of this number are represented by points on the circle with radius 1 and center at the origin. Some of these are shown in the figure below.

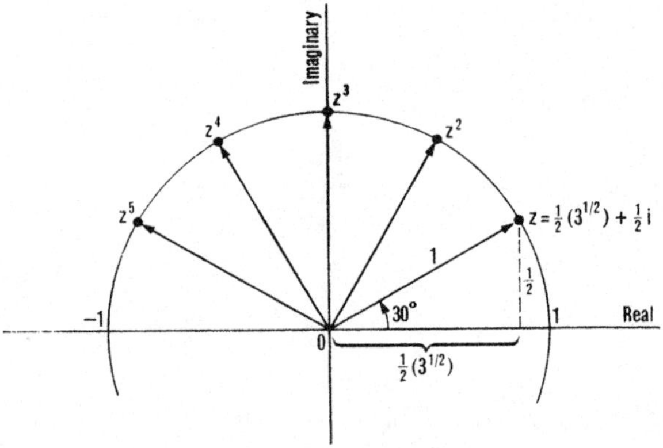

2.4.2.6. *Graphical determination of roots*

By simply reversing the construction principle described in the preceding section for determining a positive integral power of a complex number we deduce the following principle for determining the roots of a complex number.

If the complex number $a + bi$ is represented by the point N where ON has length L and inclination angle A, then an mth root

of $a + bi$ is represented by a point R such that OR has length $L^{1/m}$ and inclination angle A/m. Other mth roots of $a + bi$ will have the same vector-length $L^{1/m}$ and will have the inclination angles $(A + 360°)/m$, $(A + 2 \cdot 360°)/m$, $(A + 3 \cdot 360°)/m$, and so on to $(A + (m - 1) \cdot 360°)/m$. This gives m distinct mth roots of $a + bi$. [Note that if we were to consider $(A + m \cdot 360°)/m$ [that is, $(A/m) + 360°$)] this would lead to the same point on the graph as the one with inclination angle (A/m). Similarly, $(A + (m + 1) \cdot 360°)/m$ would duplicate the root given by $(A + 360°)/m$, etc.]

EXAMPLE

The fifth roots of $-\frac{1}{2}(3^{1/2}) + \frac{1}{2}i$.

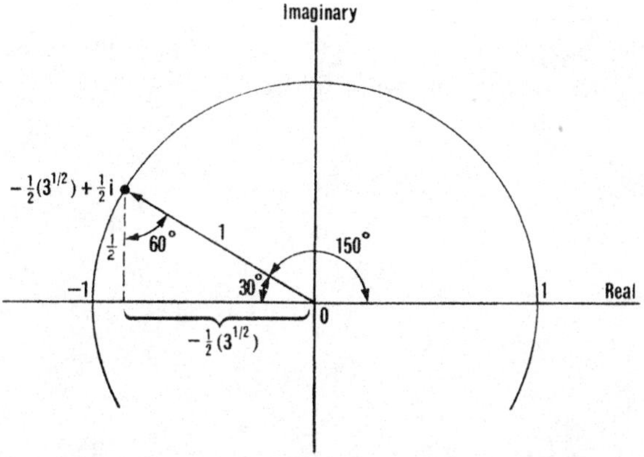

The vector-length to the point representing $-\frac{1}{2}(3^{1/2}) + \frac{1}{2}i$ is

$$\{[-\tfrac{1}{2}(3^{1/2})]^2 + (\tfrac{1}{2})^2\}^{1/2} = \{\tfrac{3}{4} + \tfrac{1}{4}\}^{1/2} = 1^{1/2} = 1.$$

Hence the vector-length for each of the fifth roots of $-\frac{1}{2}(3^{1/2}) + \frac{1}{2}i$ is $1^{1/5}$, or 1. The angle of inclination of the vector drawn from the origin to the point representing $-\frac{1}{2}(3^{1/2}) + \frac{1}{2}i$ is 150°. Hence the angles of inclination for the vectors drawn to the various fifth roots are:

$$\frac{150°}{5} = 30°$$

$$\frac{150° + 360°}{5} = \frac{510°}{5} = 102°$$

$$\frac{150° + 2 \cdot 360°}{5} = \frac{870°}{5} = 174°$$

$$\frac{150° + 3 \cdot 360°}{5} = \frac{1230°}{5} = 246°$$

$$\frac{150° + 4 \cdot 360°}{5} = \frac{1590°}{5} = 318°.$$

Hence the five fifth roots of $-\frac{1}{2}(3^{1/2}) + \frac{1}{2}i$ are represented by the points shown in the following figure.

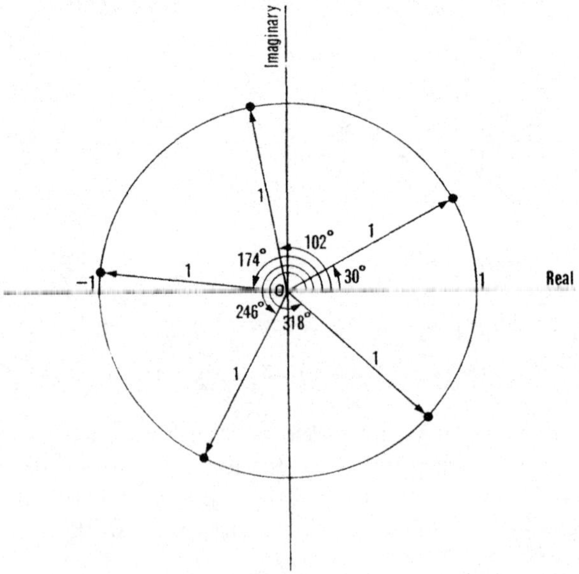

EXERCISES

1. First perform the indicated operations graphically. Then check by using the standard formulas (for sum, difference, product and quotient) and comparing results.

(a) $(8 - i) + (2 + 3i)$ (b) $(8 - i) - (2 + 3i)$

(c) $(8 - i) \cdot (2 + 3i)$ (d) $(8 - i) / (2 + 3i)$.

2. Compute all the positive integral powers of $z = \frac{1}{2}(3^{1/2}) + \frac{1}{2}i$ up to and including the twelfth. Do these in order, first obtaining the second power of z by multiplying z by itself, then obtain z^3 by multiplying the value found for z^2 by z, them multiply z^3 by z to get z^4, etc. Show all the results on a graph. How do the results compare with the figure on page 256?

3. In the manner of Exercise 2, find all positive integral powers of $\frac{1}{2}(2^{1/2}) + \frac{1}{2}(2^{1/2})i$ up to and including the eighth. Show them on a graph.

4. Find all the fourth roots of i and show them on a graph.

5. Find the value of i^{1000} and explain your method.

6. Find the two square roots of $3 + 4i$ without the use of a graph. Explain your method.

7. From the dimensions shown in the figure on page 258 the actual value of the first fifth root of $-\frac{1}{2}(3^{1/2}) + \frac{1}{2}i$ can be estimated as $.87 + .50i$. By inspection of this figure obtain a similar estimate for each of the other four fifth roots of $-\frac{1}{2}(3^{1/2}) + \frac{1}{2}i$. (Give results to two decimal places.)

INDEX

ELEMENTARY VECTORS

E. Œ. WOLSTENHOLME, *Principal Lecturer in Mathematics, Sheffield City Training College*

This book provides an introductory course on vector analysis. The treatment is rigorous but set out in sufficient detail for a beginner to grasp both the nature of a vector and the fundamental processes of vector manipulation—addition, subtraction, multiplication, differentiation, and integration. Emphasis is placed on the structure of a system of vectors, noting its analogies with, and differences from real and complex algebra. Vector methods are used in the solution of problems in the concrete world of geometry and mechanics, and the way is prepared for further reading about field vectors and vector spaces. The book will provide a useful introduction to the subject of vectors for university students, whilst the detailed treatment of fundamental principles makes it equally suitable for students in training colleges and colleges of technology.

MODERN MATHEMATICS IN SECONDARY SCHOOLS

D. T. E. MARJORAM, B.Sc., Dip.Ed., A.F.I.M.A.,
Deputy Headmaster,
Doncaster Technical Grammar School for Boys

In providing a simple introduction to most of the topics which have found, or are finding, their way into revised syllabuses in school mathematics—and in discussing ways in which these subjects might be treated in the classroom—the author takes a completely new look at familiar concepts. Topics covered—most of which are new to Secondary School mathematics syllabuses—include sets, applications of set theory and Boolean algebra, groups, matrices, vectors, linear programming, statistics and patterns in arithmetic. Each topic is developed out of preceding ones and numerous worked examples are given. Each chapter concludes with a list of the latest relevant publications for further reading, and the appendix contains a variety of exercises on the whole book together with answers, hints and, in many cases, full solutions.
The book will be of great value to teachers in that it will undoubtedly help those faced with the task of preparing pupils for any of the new syllabuses in 'O' level mathematics. Teachers in training will also find it useful, and it may be used by sixth form pupils as supplementary reading.

Printed in Dunstable, United Kingdom

91338438R00157